Studies in Russia and East Europe

This series includes books on general, political, historical, economic and cultural themes relating to Russia and East Europe written or edited by members of the School of Slavonic and East European Studies, University College London, or by authors working in association with the School.

Titles include:

Roger Bartlett and Karen Schönwälder (*editors*)
THE GERMAN LANDS AND EASTERN EUROPE

John Channon (*editor*)
POLITICS, SOCIETY AND STALINISM IN THE USSR

Stanislaw Eile
LITERATURE AND NATIONALISM IN PARTITIONED POLAND, 1795–1918

Rebecca Haynes
ROMANIAN POLICY TOWARDS GERMANY, 1936–40

Geoffrey Hosking and Robert Service (*editors*)
RUSSIAN NATIONALISM, PAST AND PRESENT

Krystyna Iglicka and Keith Sword (*editors*)
THE CHALLENGE OF EAST–WEST MIGRATION FOR POLAND

Andres Kasekamp
THE RADICAL RIGHT IN INTERWAR ESTONIA

Stephen Lovell
THE RUSSIAN READING REVOLUTION

Marja Nissinen
LATVIA'S TRANSITION TO A MARKET ECONOMY

Danuta Paszyn
THE SOVIET ATTITUDE TO POLITICAL AND SOCIAL CHANGE IN CENTRAL AMERICA, 1979–90

Alan Smith
THE RETURN TO EUROPE
The Reintegration of Eastern Europe into the European Economy

Jeremy Smith
THE BOLSHEVIKS AND THE NATIONAL QUESTION, 1917–23

Jeanne Sutherland
SCHOOLING IN THE NEW RUSSIA

Studies in Russia and East Europe
Series Standing Order ISBN 0–333–71018-5
(*outside North America only*)

You can receive future titles in this series as they are published by placing a standing order. Please contact your bookseller or, in case of difficulty, write to us at the address below with your name and address, the title of the series and the ISBN quoted above.

Customer Services Department, Macmillan Distribution Ltd, Houndmills, Basingstoke, Hampshire RG21 6XS, England

The Return to Europe

The Reintegration of Eastern Europe into the European Economy

Alan Smith

Reader in East European Economics
School of Slavonic and East European Studies
University College London

in association with
SCHOOL OF SLAVONIC AND EAST EUROPEAN STUDIES
UNIVERSITY COLLEGE LONDON

Published by PALGRAVE
Houndmills, Basingstoke, Hampshire RG21 6XS and
175 Fifth Avenue, New York, N. Y. 10010
Companies and representatives throughout the world

PALGRAVE is the new global academic imprint of
St. Martin's Press LLC Scholarly and Reference Division and
Palgrave Publishers Ltd (formerly Macmillan Press Ltd).

Outside North America
ISBN 0–333–74045–9

In North America
ISBN 0–312–23262–4

This book is printed on paper suitable for recycling and
made from fully managed and sustained forest sources.

A catalogue record for this book is available from the British Library.

Library of Congress Catalog Card Number: 99–087195

Transferred to digital printing 2003

Printed and bound in Great Britain by
Antony Rowe Ltd, Chippenham and Eastbourne

To Ruth and Steven

Contents

List of Tables

Preface

The purpose of this book is to examine the ability of the central and south-east European economies to withstand competitive pressures on entry to the EU. These economies are still experiencing structural problems inherited from over-concentration on heavy industrial production under communism. A major gap still exists between the income levels and economic structures of these states and the existing members of the EU. The gap in income levels is substantially larger than on the occasion of earlier enlargements of the EU to admit Greece, Portugal and Spain. Unlike trade relations between existing members of the EU, where intra-industry trade predominates, trade relations between central and eastern Europe and the EU largely involve the exchange of labour-intensive goods produced in eastern Europe for human capital-intensive goods produced in the EU. The export structure of some central European economies (notably Hungary, Slovenia and the Czech Republic, and to a lesser extent Poland) which have attracted foreign direct investment is improving and contains a higher proportion of human capital-intensive products. At the other end of the spectrum, Romania and Bulgaria have failed to attract significant inflows of foreign direct investment and have become increasingly dependent on exports of labour-intensive goods. This raises the possibility that the central European economies will benefit from 'agglomeration' and 'first mover' effects which will denied to the Balkan economies. These problems could be exacerbated by a phased accession in which the poorer European economies fail to attract private investment and do not receive equivalent transfers from EU budgets. The isolation of the Balkan economies has been increased by the conflicts in former Yugoslavia. This raises fundamental questions about the nature of economic development in Europe in the post-communist era which has emerged, in which some regions of Europe fail to attract investment and become dependent on the export of low-wage goods with the possibility of widespread poverty.

This book is part of a research project conducted at the London Business School and the School of Slavonic and East European Studies, University College London with funding from the ESRC under its 'One Europe or Several?' programme. I am grateful for advice and assistance from David Dyker, Saul Estrin, Tomasz Mickiewicz, Mario Nuti, Peter Westin, Milica Uvalic, Urmas Varblane and Marzenna Weresa. I am especially grateful to Mark Otto, who financed the acquisition of the COMEXT database by the library at the School of Slavonic and East European Studies.

Finally, I would like to express my appreciation to my wife, Ruth, and son, Steven, for their support and forbearance, particularly in the latter stages of writing.

Alan Smith

List of Abbreviations

CEFTA : Central European Free Trade Area
CIS : Commonwealth of Independent States
CMEA : the Council for Mutual Economic Assistance, the trade grouping of the former centrally-planned economies; also known as Comecon
CMEA-5 : Bulgaria, Czechoslovakia, Hungary, Poland and Romania
EBRD : European Bank for Reconstruction and Development
EU-12 : the twelve members of the European Union before the enlargement of 1 January 1995 to admit Austria, Finland and Sweden.
EU-15 : the fifteen memers of the European Union as at 1 January 1995
ESI : Export Specialisation Index
FDI : Foreign Direct Investment
GDP : Gross Domestic Products
IBEC : International Bank for Economic Cooperation (the CMEA clearing bank)
IMF : International Monetary Fund
OECD : Organization for Economic Cooperation and Development
RCA : index of revealed comparative advantage
SITC : United Nations Standard International Trade Classification
TR : Transferable Rouble, the unit of account for CMEA foreign trade
UNECE : United National Economic Commission for Europe

1
The Background to Eastward Enlargement

1.1 The Soviet trading bloc and its dissolution

The collapse of communism in central and south-eastern Europe and the dissolution of the Soviet Union in December 1991 resulted in the breakup of the one of the largest trade preference zones in the world. The zone extended over a fifth of the world's land surface and embraced a population of 400 million people who were composed of more than 100 nationalities. In 1989 the configuration of this zone, which was known as the Council for Mutual Economic Assistance (CMEA), or more popularly as Comecon, was relatively simple. The CMEA was dominated, economically and politically, by the Soviet Union, which itself consisted of fifteen nominally independent republics, of whom the Russian Republic was by far the most important in terms of population and resources. The other members included the five central and south-east European states which fell under Soviet hegemony after the second world war but who retained their own separate national institutions and identities – Bulgaria, Czechoslovakia, Hungary, Poland and Romania plus the German Democratic Republic which consisted of the eastern Lander of Germany which were reconstituted into a separate independent republic after falling under Soviet occupation. Three other non-European states, Cuba, Mongolia and Vietnam were also full members of CMEA and two other European socialist republics, Albania and Yugoslavia, were associate members of the CMEA but played only a minor role in its activities.

The majority of these states underwent a complex political and economic reconfiguration in the early 1990s. The dissolution of the Soviet Union in December 1991 resulted in the emergence of fifteen separate independent republics, twelve of whom are currently members of a far looser political and economic organisation, the Commonwealth of Independent States (CIS).[1] The three Baltic States, Estonia, Latvia and Lithuania, declared their independence in 1990 and declined to join the CIS after the collapse of the Soviet Union.[2] The German Democratic Republic was re-unified with the Federal Republic of Germany in 1991 and effectively became the first former socialist economy to be become a full member of the European Union. Albania, Bulgaria, Hungary, Poland and Romania retained their independent identities and the territories they had inherited as a result of the post-Second World War settlement in Europe. Czechoslovakia briefly became the Czech and Slovak Federative Republic before the Czech Republic and Slovakia divided into separate republics in January 1993. The former Socialist Republic of Yugoslavia has been riven by civil conflict since 1991 and by the middle of 1999 had been divided into Bosnia-Hercegovina, Croatia, Slovenia, the former Yugoslav Republic of Macedonia, and the Federal Republic of Yugoslavia (comprised of Montenegro and Serbia). Serbia has itself been riven by the civil war in Kossovo which resulted in the conflict with NATO in 1999.

1.2   Negotiations for enlargement

For the majority of the central and south-east European states the concept of a 'return to Europe', including membership of European economic and political institutions and the redirection of their trade flows away from the former Soviet Union towards western Europe became a major political, as well as an economic, objective. This clearly implied stronger relations with the EU and the goal of eventual membership. The first steps in this process were the conclusion of association agreements in 1991 between Czechoslovakia, Hungary and Poland and the EU in 1991 which were followed by agreements with Bulgaria and Romania in 1992.[3] The association agreements allowed for the eventual creation of a free trade area between the EU and the east European states in industrial products over a ten year period. The agreements provided some initial protection to the transition economies by permitting them to reduce their tariffs on EU products more slowly than the EU dismantled its restrictions on imports from the east European states. The two critical weaknesses in the agreements from the perspective of the east European states were that sensitive goods[4] and agricultural products were excluded from the commitment to free trade.

   The EU formally committed itself to the concept of enlargement to include former communist economies at the meeting of the EU Council

Copenhagen in June 1993. The European Council established that the central and east European economies with association agreements would be allowed to become full members of the EU, if they wished to, provided that they could satisfy the economic and political criteria for membership. These criteria included the establishment of a pluralist democracy with full respect for human rights and the protection of minorities; the creation of a functioning market economy and the ability to cope with competitive pressures within the EU; the ability to undertake the full responsibilities of EU membership including the free circulation of goods and services, capital and labour and adherence to the aims of political, economic and monetary union. The declaration by the European Council also stated that enlargement would also be contingent on the ability of the EU to absorb new members without threatening the momentum towards European integration while the new members would be required to accept and implement EU decisions (Grabbe and Hughes, 1997; Avery and Cameron, 1998).

The Copenhagen Council did not establish a timetable for enlargement and consequently did not provide a clear indication of the phasing of enlargement. In December 1994, the Essen Council approved a detailed pre-accession strategy drawn up by the European Commission and included the three Baltic states (Estonia, Latvia and Lithuania) and Slovenia in the pre-accession process. This brought the number of potential central and south-east European applicants to ten (including the Czech Republic and Slovakia as separate states). Each of the ten states made a formal application for membership between 31 March 1994 (Hungary) and 10 June 1996 (Slovenia). Other former Soviet and Yugoslav republics which had signed either Trade and Cooperation Agreements or European Partnership Agreements with the EU have not received an assurance of eventual membership or indication of when negotiations for entry might commence. The following meetings of the European Council in Cannes and Madrid in 1995 and Florence and Dublin in 1996 made relatively slow progress toward the principle of eastward enlargement. The Madrid Council instructed the Commission to draw up Opinions on the east European applications for membership. In July 1997 the European Commission decided that five of the ten applicant states had made sufficient progress towards the fulfilment of the conditions established at the Copenhagen summit that they should be allowed to enter negotiations for accession. The five selected states consisted of three central-east European republics that had formerly been members of the CMEA (Czech Republic, Hungary, and Poland), one Baltic State (Estonia) and a former Yugoslav republic (Slovenia). The

Commission concluded that negotiations for accession for the remaining five central and south-east European states would commence once the state concerned had made sufficient progress to satisfy the conditions of membership established by the Copenhagen Council. Slovakia, despite having the third, or fourth, highest per capita income of the ten applicant states (see Table 1.1) was omitted from negotiations for entry on the grounds that it did not meet the political criteria for entry established at Copenhagen. The inability to satisfy the economic criteria for membership were central to the Commission's decision not to open immediate negotiations for entry with the other two Baltic states (Latvia and Lithuania) and the Balkan states (Bulgaria and Romania). The Luxembourg Council of December 1997 ratified the Commission's recommendations to open negotiations for accession with the named five central-eastern European economies and the detailed negotiations commenced on a bilateral basis in 1998.

As the European Commission effectively divided the applicant states into two distinct groups for purposes of negotiations, the applicants will be referred to as 'first tier' and 'second tier' according to the Commission's decisions, for purposes of simplicity. However, the number of rounds of enlargement and the number of participants who might be included at any stage of enlargement remains fluid. No date for accession had been determined by the start of entry negotiations at the beginning of 1998. Although there was an initial expectation that a first round of enlargement, involving some or all of the 'first tier' states would take place at some time between 2002 and 2005, it is possible that it could be delayed beyond that period. A second problem is that the identity of the candidates that will be included in the first and subsequent tiers of enlargement has not been finally decided. Each of the candidates for accession conducts its negotiations with the EU separately, and it is by no means certain that either the Commission will be satisfied that all five candidates for early entry will be capable of meeting all the required political and economic criteria for entry simultaneously or that the Council, which has to unanimously approve accession treaties, will give its assent. Furthermore, the entry of each of the candidates has different implications for the EU itself, ranging from pressures on the EU budget to questions of policing EU borders and the division of ethnic groups and families inside and outside the EU. This raises the possibility that the first round of enlargement will not include all five of the potential 'first-tier' entrants or that the first round of enlargement will be delayed to permit the slowest entrants to adjust. A major delay could also result in further redistribution of the candidates between the three different categories negotiating with the EU. However, given the size of the ten applicants (with a population of 105.3 million) and the different sets of problems that each poses for the EU it is highly improbable that all ten will be admitted simultaneously and a phased enlargement in which some candidates are admitted, whilst others remain with some from of transitional status appears inevitable.

1.3   A difficult enlargement?

Eatwell at al (1997) provide four cogent arguments to indicate that the accession of the central and south-east European economies to the EU will differ substantially from the four earlier European enlargements.[5] Firstly, they argue, that the gap in economic development between the transition economies and the existing members of the EU is significantly greater than was the case in previous enlargements. Indeed, the enlargements 1973 to admit Denmark, Ireland and the UK and in 1995 to admit Austria, Finland and Sweden, primarily involved the accession of states with per-capita income levels close to, or above, the average for the EU. Secondly, the EU itself has evolved from a customs union with a common external tariff at the time of the first enlargement in 1973 to a single market involving free mobility of goods, services and factors of production in 1992 and will have undergone a further substantial deepening of economic integration involving at least the preliminary steps to monetary union by the time the transition economies will be accepted for membership. Thirdly, it is possible that a phased accession with substantial delays between the various tiers of enlargement will create and institutionalise new economic and political divisions within Europe. Although the redrawing of EU boundaries will unite some ethnic groupings that have hitherto been divided, it will divide others who in some cases have only recently acquired, and become accustomed to, the benefits of greater cross-border mobility.[6] Similarly the redefinition of EU borders will have an adverse effect on cross-border trade between former-socialist economies that has burgeoned since the collapse of central planning. As the EU becomes more inclusive, the trading opportunities for those economies that are excluded from the various tiers of enlargement will become increasingly limited if the EU is unable, or unwilling, to provide equal trade opportunities to the excluded states.

Finally, the four previous enlargements involved the accession of existing market economies with existing market institutions who possessed a structure of investment that had been largely determined by market forces. The trade structure of these state before accession had been largely determined by the principles of comparative advantage and was directed towards market economies and the EU in particular.[7] Manufacturers in these countries were already in direct contact with the demands of EU markets and had long experience of production according to the quality standards and technical specifications required in EU markets. The transition economies are emerging from a situation in which trade patterns were largely determined by the pattern of investment and industrial development which was first adopted in the Soviet Union in the 1930s and was subsequently imposed on the east European economies in the late 1940s and early 1950s. This resulted in the construction of a capital stock which reflected Stalin's preference for heavy engineering industry and which was geared towards meeting regional (internal) demand rather than the demands of the world market. Even as late as 1989, the structure of trade relations of the socialist states was influenced by the Stalinist concept of 'two world economic systems'. This required the socialist states inside and outside the Soviet Union to conduct trade between themselves at the expense of the pursuit of trade flows with non-socialist countries that might have been regarded as more rational from a microeconomic, geographical, or even historical perspective. As a result the CMEA states were responsible for less than 3 per cent of all international trade flows that were conducted outside the CMEA itself in 1989.[8]

1.4   Relative income levels and living standards in the EU and the transition economies

How backward are the transition economies in relation to the current fifteen members of the EU? Table 1.1 provides some basic socio-economic indicators for the 10 transition economies with Europe agreements, a number of other European transition economies and selected EU economies in order to permit a preliminary assessment of their relative income levels and living standards. World Bank estimates of income on a purchasing power parity basis have been derived by converting GDP per capita in the domestic currencies of the transition economies into US$ by means of an exchange rate which indicates the ratio of the number of units of the domestic currency which would be required to buy a given amount of goods and services in that country to the number of dollars that would be required to purchase the same amount of goods and services in the United States.[9] These have then been converted into ECU at the 1997 exchange rate of the ECU 1.00:$1.134 to arrive at the figures in column three. Estimates of per-capita GDP on an exchange rate basis in column four have been estimated by converting GDP data from domestic currencies into ECU at the prevailing exchange rate for 1997. Table 1.1 indicates that the average per-capita income, estimated on the basis of market exchange rates of the five economies which are negotiating for accession in the first tier of enlargement came to ECU 3685 in 1997 and was only 18.3 per cent of the level of the average of the EU-15 of ECU 20,109. Each of the transition economies shown in Table 1.1 has a per capita income below that of the poorest EU economy (Portugal) according to both market exchange rate and purchasing power parity estimates in 1997, although this reflects a major improvement in estimated income levels in Greece since the early 1990s. Slovenia, the richest economy in central and south-east Europe with a per capita income of ECU 8536 on a market exchange rate basis ranked 27th in the world after Portugal and Greece at 26th and 24th respectively. Slovenia, the Czech Republic and Hungary have been classified by the World Bank as upper-middle income economies. All other transition economies negotiating for entry into the EU are categorised as 'lower middle income economies' (World Bank, 1996) The average income of the economies that were not asked to open immediate negotiations for accession estimated on the basis of market exchange rates at ECU 1508 was only 7.5 per cent of the EU average, with Bulgaria's income estimated at only 5.0 per cent of the EU average and 10.9 per cent of that of Portugal.

Although income comparisons based on market exchange rates are frequently quoted for political purposes, they provide a substantial over-estimate of the gap in real incomes between the transition economies and the existing members of the EU and estimates of the time it will take to close the gap to any appreciable degree. Estimates of incomes derived from purchasing power parity rates provide a more realistic comparison of the real incomes of citizens in different countries, and hence their living standards, than estimates based on market exchange rates. The divergence between estimates of income derived from purchasing power parity rates and incomes estimated on the basis of market-determined exchange rates is generally largest in relatively poor countries with low wage rates as the wage-cost element of goods and services which are sold exclusively, or predominantly, in domestic markets (non-tradeables) is lower in low-income countries than in high-wage economies. This has been augmented, in the case of the central and south-east European economies, by the adoption of relatively depreciated exchange rates in the initial stage of the transition to market economy to provide some price protection to domestic industries and to make exports more

Table 1.1　Basic socioeconomic indicators of transition economies and selected EU economies in 1997

	Population		Income per capita (ECU 000)		Infant mortality	Life expectancy[c]	
	million	% urban	ppp basis[a]	x-rate basis[b]	per 1000 live births	Male	Female
First Tier							
Czech	10.3	66	10 035	4 586	6	70	77
Estonia	1.5	74	4 417	2 936	10	63	76
Hungary	10.1	66	6 173	3 907	11	65	75
Poland	38.7	64	5 626	3 166	12	68	77
Slovenia	2.0	52	11 040	8 536	5	71	78
Weighted average		65	6 583	3 684		68	77
Second Tier							
Bulgaria	8.3	69	3 403	1 005	16	67	75
Latvia	2.5	73	3 218	2 142	16	63	76
Lithuania	3.7	73	3 977	1 966	10	65	76
Romania	22.5	57	3 783	1 252	22	65	73
Slovakia	5.3	60	6 922	3 262	11	69	77
Weighted average		62	4 086	1 571		66	74
Other Transition							
Albania	3.2	38	1 232	661	37	69	75
Croatia	4.5	57	3 376	4 065	9	68	77
FYR Macedonia	2.0	61	1 879	961	16	70	74
Moldova	4.3	53	1 824	476	20	64	71
Russia	147.2	77	3 695	2 416	17	60	73
Ukraine	50.9	71	1 913	917	17	62	73
Weighted average		74	3 168	2 012		61	73
Existing Members							
Finland	5.1	64	16 737	21 234	4	73	81
Germany	82.0	87	18 783	24 920	5	73	80
Greece	10.4	60	11 534	10 590	8	75	81
Ireland	3.6	58	14 762	16 119	5	74	89
Portugal	9.9	37	12 205	9 215	7	72	79
Spain	39.1	77	13 862	12 795	5	73	81
UK	58.4	89	18 095	18 262	6	74	80
EU-15 (weighted average)		78	17 634	20 109		74	81

Notes: a) purchasing power parity basis; b) exchange rate basis; c) life expectancy at birth.
Source: Populations from EBRD Transition Report 1998. Basic data from World Development Report 1998/9, World Bank, 1999. Per-capita incomes converted into ECU at ECU = $1.134. Weighted averages are author estimates.

competitive in western markets. Consequently a relatively low domestic wage measured in ECUs will buy more goods and services in a poorer central and south-east European economy than the same wage, measured in ECUs, will in a richer European country.

Conventional economic theory predicts that the divergence between income measured on an exchange rate basis and a purchasing power parity basis will narrow as the transition economies become more competitive in EU markets, resulting in a real appreciation of the exchange rate. As the barriers to factor mobility are removed, labour will be expected to move to higher-wage regions and capital will seek out lower-wage, but skilled labour, reducing income differentials and closing the gap real living standards between the transition economies and the EU (see Chapter 2). However, this process is far from complete amongst the existing members of the EU. It is also noticeable from Table 1.1 that the incomes of the poorer EU economies (Spain, Portugal and Greece) measured on a purchasing power parity basis are higher than incomes measured on an exchange rate basis, while income estimates based on purchasing power parity for the richer EU states are actually lower than estimates based on market exchange rates. Countries at intermediate levels of income (Italy, Ireland and the UK) have exchange-rate based incomes that are slightly higher than purchasing power parity incomes. This implies that European exchange rates were overvalued in relation to the US$ in terms of purchasing power parity in the period 1995–7 and that estimates based on market-exchange rates give a falsely high indication of real living standards in the richer EU countries during this period.

The gap between income levels in the transition economies and the EU economies in Table 1 narrows considerably when estimates based on purchasing power parity are considered, but still remains significant. The average per-capita income for the 'first-tier' economies, estimated on a purchasing power parity basis, rose to ECU 6583 in 1997 but was still only 37.3 per cent of the average for the EU-15 of ECU 17,634, while the average income of the 'second-tier' economies at ECU 4027 was only 22.8 per cent of the EU average.

The average per-capita income of the richest transition economy, Slovenia at ECU 11,040 remained just below that of the poorest EU economy, which became Greece at ECU 11,534 when estimated on a purchasing power parity basis. Even Hungary's income was only just over half that of Greece on a purchasing power parity basis. Furthermore, the two economies with the highest per-capita incomes are newly independent states that have 'broken away' from federations which contained poorer regions which has increased mean income levels in the breakaway states. The per capita incomes of all the remaining transition economies are less than 37 per cent of the average income for the EU-15 and less than 56 per cent of the poorest member.

Table 1 also provides other socioeconomic indicators that may be considered to reflect or to determine comparative living standards. Sen makes some powerful arguments to support the use of life expectancy statistics as an indicator of the standard of living. Although life expectancy is not strongly correlated with GNP per capita per se, the factors that are related to high rates of life expectancy are facilitated by relatively high levels of GDP and also include factors that make a positive contribution to feelings of well-being. These include low rates of inequality and poverty, high absolute levels of expenditure on health care, sanitation and education and the provision of safe working and social environments. (Sen, 1995). Furthermore life-expectancy data are subject to fewer problems of statistical manipulation, definition and international comparison than other socioeconomic data. Comparisons of life-expectancy are complicated by the larger differential between female and male life expectancy in former communist economies and in the former Soviet republics in particular. Table 1.1 shows that male life expectancy at birth at 68 years for 'first-tier' economies and 66 for 'second-tier' economies is substantially lower than the EU average of 74. Female life expectancy at 77 and 74 in the first and second-tier economies respectively also fall below the EU average of 81. There has been a significant fall in (especially male) life expectancy since the collapse of communism in the former Soviet Republics and to a lesser extent in Romania and Bulgaria. However this phenomenon has not affected the central European republics namely, the Czech Republic, Poland, Hungary, Slovakia and Slovenia. The net effect is that life expectancy has actually increased in all the first-tier economies, except Estonia, since 1989 and has fallen in all other transition economies except Slovakia.

Similar argument to those outlined in the preceding paragraph can be made for including infant mortality as an indicator of comparative living standards. Infant mortality per 1000 live births was higher in 1997 than the EU average of six, in each transition economy except Slovenia and the Czech Republic. In the case of the other 'first tier' economies, infant mortality was nearly double the EU average. All of the 'second-tier' economies (except Slovakia) have an incidence of infant mortality which is more than double that of the EU, with Romania having an infant mortality of 22 deaths per 1000 live births which was nearly four times the EU average. Although these figures are substantially poorer than those for high-income economies according to the World Bank definition, they still compare favourably with the average of 40 for all middle-income economies and 36 for other upper-middle income economies.

The data in Table 1.1 confirm that all the transition economies (with the exception of Slovenia) were substantially poorer than existing members of the EU in the mid-1990s. This still leaves the question of whether the gap between the income levels of the 'first-tier' transition economies and that of the EU is larger than that between other less-wealthy states and the EU on the occasion of previous enlargements. The most relevant comparison is the enlargement of the EU to include Spain and Portugal in 1986. Table 1.2 uses World Bank data to provide a comparison of purchasing power parity estimates of the incomes of Spain and Portugal with that of the UK in 1987 which has been taken as a proxy for the average level of income in the EU. These figures indicate that per-capita incomes in Spain and Portugal were 71 per cent and 58.5 per cent respectively of the UK level in 1987. The other two less-wealthy members of the EU, Ireland and Greece had income levels of 57.4 per cent and 59.5 per cent of the UK average respectively. Again, only Slovenia and the Czech Republic with purchasing power parity incomes of 62.6 per cent and 56.9 per cent of the EU average in 1997 had a relative income gap that was comparable with that of Portugal, Greece and Ireland in 1987.

What are the prospects for a major closure of the income gap between the central and east European economies and the EU in the period covering their entry into the EU?. If the transition economies were to sustain annual growth rates of per capita incomes measured on purchasing power parity basis of just over 7 per cent per annum, their incomes

Table 1.2   Relative income levels of new entrants to EU on a purchasing power parity basis.

Purchasing power parity income per capita as % of UK level in 1987		Purchasing power parity income per capita as % EU average in 1997	
Spain	71.0	Slovenia	62.6
Portugal	58.5	Czech	56.9
Greece	59.5	Hungary	35.0
Ireland	57.4	Poland	31.9
		Estonia	25.0

Source: Estimated from data in World Development Report, 1996 and 1999.

would double over a ten-year period. If the average growth of income in the EU grew by only 2.5 per cent per annum, average income in the EU on a purchasing power parity basis would grow to ECU 22,500 over the same period. On this basis, per-capita incomes in Slovenia and the Czech Republic would be approaching the EU average. Of the remaining 'first tier' candidates Hungary would have a per capita income 55 per cent of the EU average, Poland of 50 per cent and Estonia of just under 40 per cent. With the exception of Slovakia, incomes of economies in the 'second tier' would be between 30 and 40 per cent of the EU average in 2007. These estimates are very crude. However the growth rates have been chosen to maximise the potential for closing the income gap. Estimates of average incomes measured on a purchasing power parity basis in 1994 and 1997 indicate that they grew by 3.7 per cent per annum in the EU-15 over the period, by 5.2 per cent in the first tier economies, but actually fell by 0.1 per cent in the economies in the second tier as a result of major falls in output in Romania and Bulgaria over this period. Incomes in Spain, Portugal and Greece grew at about the same rate as those of the economies in the first tier of negotiations. It must be admitted that estimates of income based on purchasing power parity are subject to error and the differences in growth rates reflect differences in the methods of estimation over the period. Nevertheless the estimates above give an indication of the scale of the problems that must be overcome to generate a real convergence of incomes between the existing and potential members of the EU.

1.5   Directions of trade of the central and south-east European economies

1.5.1   Patterns of trade in the inter-war years and the early post-war period

The concentration of trade of the central and south-east European economies on the Soviet Union and other members of the CMEA in the late 1940s and early 1950s, had involved a substantial change in the direction of trade away from their traditional trade partners in the inter-war period when the trade flows of the central and south-east European states had been concentrated on central Europe (see Table 1.3). Exports to the Soviet Union were relatively unimportant for the central and south-east European economies in 1928 (the year before the central planning was initiated in the Soviet Union) when they accounted for less than 1 per cent of the total exports of Hungary, Poland,

Table 1.3   Direction of exports of central and south-east European states in 1928 (per cent)

	Europe	Germany	Austria	eastern Europe	Soviet Union
Poland	95.4	34.7	12.3	15.5	1.7
Czechoslovakia	86.0	26.7	14.7	16.1	1.3
Hungary	97.5	11.9	34.0	27.0	0.4
Romania	92.3	24.9	11.4	18.7	0
Bulgaria	97.4	27.6	14.5	10.5	0
Estonia	91.4	25.9	0	0	5.2
Latvia	87.1	27.1	0	3.5	9.4
Lithuania	95.3	58.1	0	0	2.3

Notes: Eastern Europe consists of Poland, Czechoslovakia, Hungary, Romania and Bulgaria.
Europe excludes the Soviet Union.
Source: The Network of World Trade, League of Nations, Geneva, 1942.

Czechoslovakia, Hungary and Romania combined. By 1938, Czechoslovakia was the only central and east European country, other than the Baltic states, to conduct trade with the Soviet Union. Germany and Austria were the most important trade partners for these economies, accounting for approximately 25 per cent and 15 per cent of total central and south-east European exports respectively. Germany was the most important market for all countries except Hungary (for whom Austria accounted for 34 per cent of exports) and Estonia (for whom the United Kingdom accounted for 34 per cent of exports). Trade between the east European economies themselves accounted for approximately 18 per cent of their total trade turnover in 1928. The depression, which resulted in a halving in world trade in the 1930s, contributed to a reduction in central and east European trade flows which was particularly severe in Czechoslovakia.

It can be argued that inter-war patterns of trade do not provide a good guide to the patterns of trade that would have developed in the post-war period without Soviet intervention. The post-war collapse of the German and Austrian economies would have greatly complicated the resumption of trade flows between them and their traditional central European partners . The central and south-east European economies were themselves in a state of collapse and the west European economies were reluctant to give assistance or credits for economic reconstruction in Hungary, Romania and Bulgaria while questions of reparations concerning their role in the war remained to be settled. As a result, the central and south-east European economies had little alternative to trade with the Soviet Union in the immediate post-war period. Nevertheless, the central and south-east European economies made several attempts to develop trade relations and to improve economic cooperation between themselves with the result that by 1948 trade between the central and south-east European economies themselves was more important than trade between central and south-east Europe and the Soviet Union (Berend, 1971, 10). However, the division of Europe into two economic blocs was accelerated by Stalin's veto of east European participation in the Marshall Plan for European Reconstruction which was announced on 5 June 1947. This was followed by the US-imposed embargo on exports of strategic and high technology goods to communist economies from recipients of Marshall Aid in western Europe (van Brabant, 1980). The slide into cold war following the communist takeover in eastern Europe in 1948, the creation of the CMEA in 1949 and the outbreak of the Korean War meant that by Stalin's death in 1953, Europe was divided into two separate economic camps.

1.5.2   Directions of trade in the 1990s and the potential for trade redirection

Is it possible to predict what would have been a more normal choice of trade partners for the central and east European economies if this politically-inspired separation had not taken place? Gravity models are a major method used by economists to predict volumes of trade between any two countries or regions. The principle behind gravity models is that trade is affected by the pull of specific factors that overcome transactions costs associated with international trade, while factors that reduce transactions costs will also stimulate trade. These include costs associated with discovering new markets and sources of supply, costs of transportation, costs of exchanging currencies and ensuring payment, costs of uncertainty surrounding currency movements and costs associated with compliance with customs regulations and environmental and other regulations in partner countries.[10]

The major factors that are considered to determine trade levels between two countries (or regions) are levels of aggregate GDP, GDP per-capita, distance and country size. A higher level of aggregate GDP in the importing country, other things being equal, will generate a greater demand for imported goods. Similarly a high level of aggregate GDP in the exporting country implies a greater potential supply of exports. However, this will be partially offset by the ability of large countries to satisfy domestic demand from domestic sources of supply, or to find domestic markets for output without incurring the transactions costs associated with international trade. In addition, country size, in the form of land mass, increases the transactions costs involved in exporting and importing by increasing the costs of transport to national borders from the point of production and from national borders to the point of consumption and will exert a negative effect on trade. Consequently large countries tend to have lower levels of trade in relation to GDP than small countries. Richer countries also tend to trade more than poorer countries, other things being equal, partly as consumers with high incomes seek out new and exotic sources of goods. Consequently, the level of GDP per capita incorporates both the high level of aggregate demand in the economy and compensates for the negative effects associated with country size and tends to be linked to high volumes of trade.[11] The distance between countries increases costs of transportation and reduces the potential for trade and can increase other transactions costs, particularly if land transport involves crossing more than one boundary. Consequently, adjacency, which can also result in lower transport costs compared with domestic trade, tends to act as a stimulus to cross-border trade. Finally, the influence of historic, cultural and linguistic links should not be neglected, particularly where these help to reduce transactions costs, even if these are difficult to measure. Gravity models (based on applying parameters derived from western economies to the central and south-east European economies) broadly concluded that there was scope for a significant increase in trade flows between central and south-east Europe and the EU, although the size of the potential increase differed from study to study. Wang and Winters (1992) and Baldwin (1994), estimated that the central and south-east European economies would be expected to double their level of exports to the EU from the levels pertaining at the end of the communist era under more normal circumstances. Brenton and Gros (1995) however concluded that the potential for a major redirection of trade had been exhausted by 1992. A study by Sheets and Boata (1996) also suggests that the potential for major trade redirection had been largely fulfilled by 1994.

1.5.3 The directions of trade in 1997

A cross-section of the directions of trade of the ten east European associated states in 1997 is shown in Table 1.4 which can examined from the perspective of the factors which determine trade flows which were outlined in the preceding paragraph. This gives an indication of the dependence of the central and east European economies on trade with the EU.[12] There are some problems of comparability between data for different counties. Trade data for the former CMEA economies and Slovenia have been taken from the IMF Direction of Trade Statistics Yearbook (DOTSY) whose primary source is data provided by national authorities. DOTSY has estimated data for Slovakia, Estonia and Lithuania from trade partners' data. Consequently estimates for the Baltic states in Tables 1.4 and 1.5 have been made directly from the foreign trade data supplied by Inter-Baltic Working Group for Foreign Trade Statistics. These data differ from those in DOTSY and tend to attach a higher value to trade with former communist economies, thereby reducing the proportion of trade conducted by the Baltic states with the EU. Czech data also contain a high volume of imports (ECU 2433 million) from European developing countries in the total that cannot be identified by trade partner although a full country nomenclature is provided.

Each of the economies shows a relatively high degree of openness to trade, (measured as the ratio of the sum of exports and imports to GDP) which is consistent with their small size. Poland and Romania, the economies with the largest populations, have the lowest degree of openness (see Table 1.5). The countries in the first tier of negotiations for enlargement, generally conducted higher volumes of trade with the EU than the countries in the second tier, according to the majority of criteria of assessment. Four of the five countries in the first tier of negotiations for enlargement had the highest share of imports from the EU in 1997, with the Czech Republic only falling below Romania as the result of the high volume of trade with Slovakia (Table 1.4). Similarly four of the five countries in the first tier had the highest share of exports to the EU, with the EU taking over 60 per cent of their exports. The exception, Estonia, had a share of 62.2 per cent according to the data provided by DOTSY, but this fell to 48.6 per cent according to national data. The share of the EU in exports and imports dropped below 50 per cent for four of the economies in the second tier of enlargement (Bulgaria, Slovakia, Latvia and Lithuania). The economies in the first tier also had the highest per-capita volumes of exports to, and imports from, the EU, with the exception of Poland This was a function of country-size as Poland had the largest absolute volume of exports to the EU.

One important feature revealed in Tables 1.4 and 1.5 is the importance of continued trade links between states that have become independent of one another since the collapse of communism. Although this partly reflects the development of regional trade patterns, the strength of the phenomenon is such that it cannot be simply explained by adjacency and proximity. In the case of Slovakia trade with the Czech Republic accounted for 25.6 per cent of exports and 24.4 per cent of imports and

Table 1.4 Direction of imports of central and south-east European economies in 1997

	Poland	Czech	Hungary	Slovakia	Slovenia	Romania	Bulgaria	Estonia	Latvia	Lithuania
Imports (ECU mn)	37 308	23 799	18 725	10 474	8 252	8 934	3 414	3 910	3 845	4 976
From EU (ECU mn)	23 811	12 364	11 674	4 774	5 564	4 690	1 428	2 313	1 723	2 138
Partners (% of total)										
EU-15	63.8	51.9	62.3	45.6	67.4	52.5	41.9	59.2	44.8	43.0
Germany	24.1	26.7	27.2	22.2	20.7	16.4	12.4	10.0	13.0	17.4
Austria	2.0	4.4	10.6	6.7	8.4	2.7	2.7	0.6	0.8	0.8
Italy	9.9	5.3	7.4	5.6	16.6	15.9	8.6	3.0	2.5	4.1
Nordic states[a]	6.7	2.8	2.9	1.7	2.1	2.0	2.6	35.1	16.6	9.7
Mediterranean[b]	2.8	1.8	1.8	1.1	2.4	2.5	5.8	0.7	0.8	1.1
Imports from EU per capita (ECU)	615	1 200	1 151	884	2 782	208	172	1 595	690	578
Former communist	15.1	21.3	18.8	47.5	16.0	21.3	37.0	23.1	47.5	42.8
CIS	8.2	7.4	10.9	17.1	3.0	15.0	30.5	17.3	31.3	30.8
Russia	6.3	6.8	9.2	14.7	2.7	12.2	26.6	14.4	19.7	25.3
CEFTA	5.9	21.2	6.8	29.6	6.6	5.7	5.3	2.4	5.0	8.7
Baltic states	0.4	0.5	0	0.1	0.1	0	0.1	3.3	11.1	3.0
Former Yugoslavia	0.6	2.3	1.0	0.7	6.3	0.6	1.1	0.1	0.1	0.3
Rest of world	21.1	26.8	18.9	6.9	16.6	26.2	21.1	17.7	7.7	14.2

Notes: a) Denmark, Finland , Sweden; b) Greece, Portugal, Spain.
Source: All countries except Baltic states estimated from data in Direction of Trade Statistics Yearbook, 1998, IMF: Geneva, 1998. Baltic states from Estonia, Latvia, Lithuania, Foreign Trade 1997, Statistical Bulletin, Riga, 1998. All data converted from $ into ECU at ECU1:$1.134.

Table 1.5 Direction of exports of central and south-east European economies in 1997

	Poland	Czech	Hungary	Slovakia	Slovenia	Romania	Bulgaria	Estonia	Latvia	Lithuania
Exports (ECU mn)	22 708	19 845	16 843	7 753	7 383	7 394	3 795	2 580	2 732	3 405
To EU (ECU mn)	14 581	11 955	11 996	3 632	4 691	4 190	1 706	1 253	975	1 107
Partners (% of total)										
EU-15	64.2	60.2	71.2	46.8	63.5	56.7	45.0	48.6	35.7	32.5
Germany	32.9	36.0	37.2	25.0	29.4	16.8	9.5	5.6	10.2	11.4
Austria	1.9	6.5	11.5	6.9	6.7	2.1	1.1	0.3	0.3	0.5
Italy	5.9	3.7	6.2	6.1	14.9	19.7	12.4	0.8	1.6	3.1
Nordic states[a]	6.6	2.0	1.7	1.1	1.5	0.8	1.1	32.4	8.9	6.1
Mediterranean[b]	1.6	1.3	2.0	1.3	1.1	3.2	12.0	0.4	0.6	1.2
Exports to EU per capita (ECU)	377	1 161	1 183	672	2 346	186	205	865	390	299
Sensitivity[c]	230	152	183	71	95	79	31	na	na	na
Former communist	24.7	28.0	19.7	46.3	27.3	11.9	24.4	42.2	54.2	60.8
CIS	15.3	4.0	7.2	6.9	4.9	6.2	17.8	26.5	40.5	46.3
Russia	8.4	3.3	5.1	3.0	3.9	3.0	7.9	18.8	24.1	24.5
CEFTA	6.8	21.2	7.6	36.1	5.5	4.6	3.1	1.1	2.6	3.3
Baltic states	1.9	0.5	0.5	0.6	0.2	0.1	0.4	14.7	10.9	11.1
Former Yugoslavia	0.6	2.3	4.2	2.8	16.6	0.9	2.5	0	0.2	0.3
Rest of World	11.1	11.8	9.1	6.9	19.9	31.4	30.6	9.2	10.1	6.7
Trade openness[d]	54.0	115.5	93.2	125.0	111.2	59.0	124.7	154.4	106.2	80.9

Notes: a) and b): See Table 1.4. c) Estimated increase in exports to EU in dollars in response to a 1 per cent increase in EU-12 GDP. Source, Feldman et al. 1998, 24. d) Sum of exports and imports as a percentage of GDP. Based on 1995–7 data. Source, Feldman et al. 1998, 17.
Source: See Table 1.4.

was ten times greater than trade with Hungary. The share of the EU in total exports and imports would rise to 63 per cent and 60.3 per cent respectively, if trade with the Czech Republic were to be discounted. Similarly the share of the EU in Czech exports and imports would rise to 69.1 per cent and 56.7 per cent if trade with Slovakia were to be omitted. The situation in the case of the Baltic states is more complicated. The substantial trade deficits of the Baltic states contribute to the paradox that trade between the Baltic states accounts for a greater share of each countries exports than of its imports. However, with the exception of Latvia, the import shares are not very large, reflecting the fact that trade between the Baltic states themselves was relatively insignificant in the inter-war years and that the trade flows of the Baltic states in the Soviet period were largely concentrated on Russia, Belarus and Ukraine. This has continued in the post-Soviet period with the CIS states, especially, Russia, Ukraine and Belarus remaining important markets for all three countries and sources of supply for Lithuania and Latvia in particular. The trade relations of the Baltic States with the EU have been concentrated on the Nordic states, Germany (which has a Baltic coastline) and to a lesser extent the UK. Finland is a major trade partners for Estonia accounting for 32.4 per cent of exports and 35.1 per cent of imports.

The most important feature of the redirection of trade for the central European economies (Poland, the Czech Republic, Hungary, Slovakia and Slovenia) has been the re-emergence of Germany as the major trade partner, both as a market for exports and as a supplier of manufactured goods. The common border has been a major factor in the development of German trade with the Czech Republic and Poland. Official trade statistics do not capture the vast amount of unrecorded 'shuttle' trade between Germany and Poland, in particular, which largely involves people travelling to Poland to buy goods in markets at prices that are considerably below those prevailing in Germany. Germany accounted for between 25 per cent (Slovakia) and 37.2 per cent (Hungary) of the recorded exports of the central European economies. Collectively, the recorded exports of the central European states to Germany at ECU 25,002 exceeded recorded imports from Germany (ECU 24,055) in 1997.

The Balkan states (Romania and Bulgaria) have been less successful in penetrating EU markets in general and the German market in particular. Exports per capita from Romania and Bulgaria to the EU were less than 20 per cent of the levels of the Czech Republic and Hungary in 1997. Romania is the only one of the ten countries seeking membership that does not share a common border, or coastline, with an EU member.

Although trade flows between Romania and Bulgaria and the EU have been disrupted by conflicts in former Yugoslavia, which have affected transit routes by land and by the river Danube, this is not the sole explanation of their relatively poor export performance in EU markets, which is examined in subsequent Chapters. There is some evidence of a regional dimension, to the trade relations of Bulgaria and Romania which uses the Black Sea as a trade route. Italy was the largest export market inside the EU (followed by Germany) for both Romania and Bulgaria. Greece is developing into a major market for Bulgarian exports, but has declined as a source of imports. The CIS market, which includes other countries bordering the Black Sea, of whom Russia is the most significant, has retained its importance for Bulgaria. Russian oil exports to Bulgaria are supplied across the Black Sea from Novorossisk to the port of Bourgas. The relatively high proportion of trade with the rest of the world reflects high levels of trade between both countries and Turkey. Romania also has a relatively high proportion of exports to the Middle East, while Korea has become an important source of imports.

1.6   Trade deficits and capital inflows

Each of the central and south-east European economies inherited problems related to indebtedness from the communist era which resulted from a combination of excess demand and the relative failure to penetrate western markets for manufactured goods. The reluctance to permit foreign investment in communist economies meant that current account deficits had to be financed by foreign borrowing (see Chapter 4). The need to modernise and recapitalise the transition economies has contributed to a renewed demand for capital inflows, the counterpart of which is a deficit on current account. Many economists had also anticipated that direct investment by multinational corporations, involving ownership and control of existing enterprises (brownfield investment) or the creation of new production facilities (greenfield investment) would become a major vehicle for the transfer of technology, modern industrial processes and managerial techniques to the transition economies. The prospect of eventual membership of the EU was expected to assist in this process by creating a regulatory framework in the central and south-east European economies that was consistent with that of the EU and by removing the barriers to EU imports of goods produced by investors in the region.

Table 1.6 provides some basic aggregated data on the size of current account deficits and net capital inflows, including foreign direct

Table 1.6 Current account deficits, indebtedness and capital inflows in 1997 (ECU million)

	Poland	Czech	Hungary	Slovak	Slovenia	Romania	Bulgaria	Estonia	Latvia	Lithuania
Current account (ECU million)										
a) 1997	-3 744	-2 783	-865	-1 322	+82	-2 070	389	-497	-304	-865
b) Cumulative 1991–7	+441	-7 760	-10 582	-3 466	+1 677	-9 915	-2 033	-1 091	-345	-2 024
Ratio current account: GDP (%)										
1995–7 annual ave.	+0.2	-5.5	-3.8	-5.3	+0.1	-6.4	+1.6	-8.6	-4.7	-9.9
Net debt at end 1997										
(ECU mn)	15 610	10 183	12 695	5 993	852	4 928	6 955	-395	-272	372
Foreign Direct Investment										
Cumulative 1989–97										
a) total (ECU mn)	10 954	7 471	14 905	1 078	1 213	2 972	1 078	891	1 198	1 121
b) per capita (ECU)	283	726	1 470	200	563	131	130	613	479	303
Inflow 1997										
a) (ECU mn)	2 684	1 124	1 852	45	283	1 079	438	113	306	192
b) as % GDP	2.2	2.4	4.6	0.3	1.8	3.5	4.8	2.7	6.3	2.3
FDI Stock 1997	14 517	5 963	14 005	1 140	2 071	2 175	832	1 012	795	918
FDI as % gross fixed capital formation										
annual ave. 1994–6	16.4	10.0	30.6	4.4	4.5	4.9	5.3	28.8	na	na
% share EU in FDI stock										
at end 1996	50	70	54	75	70	54	71	na	na	na

Sources and notes: Net debt at end 1997: Gross debt minus foreign exchange reserves estimated from Economic Survey of Europe 1998, no. 1, 153. Current account deficits for 1997, ibid. 212. Cumulative current account deficits and annual averages estimated from data in EBRD Transition Report, 1998. Foreign direct investment, cumulative 1989–97, per-capita and inflow 1997, converted from $ data in EBRD Transition Report, 1998, 81. FDI Stock and inflow as percentage GDP, fixed capital formation and EU share estimated from World Investment Report, 1988, Annexe Tables and Chap 9. All data originally in US$ converted into ECU at 1997 average exchange rate (ECU 1.0:$1.134).

investment (FDI), and its contribution to capital formation in the individual accession economies in the period since the collapse of communism. The definition of what constitutes FDI, and methods of reporting FDI, differ from source to source. Nevertheless, a fairly consistent picture emerges of the size of FDI in relation to GDP and the relative standing of the individual economies in attracting FD1. Each of the individual economies, with the exceptions of Slovenia and Poland, incurred substantial cumulative current account deficits in the period 1991–7, which were largest in absolute terms in Hungary and Romania. Hungary has attracted the highest inflow of FDI per capita in the region, which was equivalent to 30.6 per cent of gross fixed capital formation from 1994 to 1996, and Romania had little or no external debt when the Ceausescu regime was overthrown. Romania has incurred high current account deficits as a proportion of GDP in each year from 1995 to 1998. The Czech Republic has also incurred substantial current account deficits and had a particularly high deficit as a proportion of GDP in 1997, which has been roughly balanced by inflows of FDI. Poland has been more successful in attracting FDI in the mid to late 1990s after a relatively slow start. Capital inflows have helped to service Poland's debt inherited from communism and to finance current account deficits in the second half of the 1990s, after running surpluses in the mid 1990s. Current account surpluses have been combined with substantial deficits in recorded trade flows which have been offset by earnings from unrecorded trade. Although Estonia and Latvia incurred substantial current account deficits in the mid-1990s, these have been offset by capital inflows, including FDI which has enabled those economies to build up foreign exchange reserves to the extent that they exceeded external debt at the end of 1997. This is also reflected in the relatively high stock of FDI to GDP in Estonia and Latvia in 1996. Nevertheless, the size of the current account deficits reflects some problems in exporting to hard currency markets, which are more serious in the case of Lithuania. Finally, Bulgaria's trade surplus in 1997 reflects the impact of corrective measures in the wake of the financial crises of 1995–6 and difficulties in financing further current account deficits which forced Bulgaria to reduce imports in 1996 and 1997.

FDI contributed to an average of 7.5 per cent of fixed capital formation in the central and south-east European economies in the period 1993–6, but to considerably higher levels in Hungary, Poland, Estonia and the Czech Republic. The contribution of inward FDI to fixed capital formation in the other countries has been below the average for developing countries as a whole over this period and considerably below that of Latin America, China, Singapore and, Malaysia (UNCTAD, 1998). There is insufficient macroeconomic evidence to suggest that the degree of success in attracting private capital inflows has been sufficient to explain the current account deficits of the central and south-east European economies. This raises questions about the ability of the central and south-east European economies to withstand competitive pressures and expand exports to finance the imports required for the modernisation of industry, agriculture and infrastructure.

1.7 Conclusions to Chapter 1 and outline of the book

The Copenhagen principles established that the Central and East European economies (CEEs) must be capable of withstanding competitive pressures on entry to the EU. This means that producers in the CEEs must be capable, not just of competing with existing EU producers on grounds of price and quality in EU markets, but that they must also be capable of withstanding competition from existing EU producers in their domestic markets. An additional complication is that the commitment to eventual monetary union will require the new entrants to forego the use of devaluation as a measure for restoring the price-competitiveness of domestic producers under all but the most serious circumstances. The CEEs inherited an industrial structure that was created to meet the preferences of central planners, not those of the market. These preferences were enshrined in CMEA production-sharing and specialisation agreements, which created a diffused industrial structure which did not necessarily coincide with the factor endowments of the individual countries and was not based on concepts of comparative advantage. As a result, the industrial structures of the Baltic States and the central and south-east European economies that were members of the CMEA, were excessively geared to meeting the demands of the Soviet market for machinery, equipment and components for heavy engineering and ferrous and non-ferrous metallurgical industries and unsophisticated, mass-produced industrial consumer goods. The collapse of structures of demand determined by CMEA trading arrangements, implies that much of the capital stock inherited from the communist period will be ill-suited to meeting the demands of consumers determined in a market environment and that the successful reconstruction of their economies will require major inputs of capital. It will also necessitate a process of industrial rationalisation and restructuring which will involve the total closure of some industries, the concentration of production of others in fewer locations and the modernisation and possibly expansion of others.

This raises fundamental questions about which industrial sectors of the CEE economies will be able to withstand competitive pressures, and in which locations, CEE trade and economic relations with the EU are liberalised (which will involve the removal of open and hidden subsidies to loss-making industries as well as the removal of barriers to imports) as they prepare for membership. The purpose of this book is to provide some answers to these questions. This cannot be a precise undertaking. It can be argued that past provides few clues to the future and that it is impossible to predict the industrial potential of a country by examining its past industrial structures and performance. Massive inputs of capital can create new modern industrial structures that owe little to inherited structures of output. Young and well-educated labour forces are capable of undertaking retraining and acquiring the skills required to meet the demands of new industrial structures.[13] An alternative view is that successful economic development is cumulative and builds on the strengths of the past, while discarding inherited weaknesses. Under the circumstances prevailing in world markets at the end of the 1990s, major changes in the industrial structure of a country which involve international technology transfer result largely from decisions taken by multinational corporations to transfer production from one location to another, either through forms of subcontracting or through investment in new locations (Radosevic, 1999; Meyer, 1998). Private investors who have a range of global investment opportunities will only be attracted to locations and sectors that offer high potential rates of return. The decision on where to invest, or locate production, will be based on the analysis of a range of factors which will include existing infrastructure, transport costs, standards of education and training, relative wages, political stability and the economic and regulatory environment. Such decisions are seldom taken in a climate that ignores the past, and the dynamics of the recent past in particular.

This book will not concentrate on the process of foreign direct investment in CEE economies, which is the subject of other detailed studies (Estrin, Hughes and Todd, 1997; Hunya, 1997, 1998; Meyer, 1988). It will attempt to focus on the structures of trade that are emerging between the CEE economies and the EU as the CEE economies redirect their trade from the former CMEA market to the EU. The emphasis will be on the structure of CEE exports to the EU as these reflect areas in which the CEE economies have demonstrated that they are capable of competing with existing members of the EU and with competitors from outside the EU.

The fundamental economic hypotheses about the nature of trade relations within the EU and the potential impact of the liberalisation trade relations and enlargement on the CEE-economies will be examined in Chapter 2. Chapter 3 will describe the statistical tests and data sources which will be conducted in the remainder of the study. Chapter 4 will analyse the nature of CEE trade relations in the communist era and their implications for the redirection of trade to the EU. The basic features of the development of trade relations between the CEE economies and the EU since the collapse of communism will be examined in Chapter 5. Chapter 6 will compare the factor content and technological level of CEE exports to the EU with intra-EU trade and EU trade with the rest of the world. Chapter 7 will assess the areas in which the CEE economies have displayed an export specialisation and a revealed comparative advantage to assess the areas in which individual CEE economies are capable of withstanding competition from both existing EU producers and producers from outside the EU. Chapter 8 concludes by briefly examining the prospects for the successful integration of the economies of the CEE-10 into the EU.

2
International Trade Theory and EU Enlargement

2.1 The concept of international competiveness

The concept of international 'competitiveness' has aroused considerable controversy between economists and management specialists and between economists themselves.[1] The fundamental economic assumption underlying the analysis in this book is the Ricardian concept that the 'competitiveness' of a given industrial sector for any given country in international trade reflects that country's comparative, or relative, advantage in the production of the good in question, and that an economy must be competitive in the production of some set of goods and services regardless of its overall level of development. The principles of comparative advantage indicate that an economy will be competitive in the production of goods and services which it can produce relatively more cheaply (measured in terms of the foregone production of other goods and services which are used in the production of the given good) than other economies. Even if the techonological levels of a country's capital stock are so poor that it has an absolute disadvantage in the production of all goods, it must be capable of producing some goods relatively more cheaply than other economies. The critical problem that the CEE economies inherited from communism in this respect was that the structures of production that they had adopted under forced industrialisation and the Stalinist model of development had been determined by political priorities and were not based on concepts of comparative advantage. Enterprises were then provided with a guaranteed market for products which were not competitive in world markets and were protected from international competition by the state monopoly of foreign trade which only allowed external producers access to CMEA markets under strictly controlled conditions. Domestic price structures under central planning were not determined by either domestic or global supply and demand conditions and differed substantially from world market prices. Enterprises were enabled to maintain production for domestic markets in areas in which they were not competitive on world markets by a combination of subsidies, soft credits and centrally manipulated prices combined with the absence of procedures for making chronic loss-making enterprises bankrupt.

Consequently as domestic markets and foreign trade are progressively liberalised in transition economies, the relative prices of internationally tradeable goods in CEE markets will be increasingly determined by relative prices in world markets, while the absolute level of prices of tradeable goods will be determined by the exchange rate as consumers choose to import goods that are cheaper than domestically produced goods and producers seek to export goods which receive higher prices in world markets. At the same time, the price of non-tradeable goods and internationally immobile inputs has been determined to a growing degree by their relative abundance/scarcity in relation to domestic demand. Price-elasticity optimists argue that domestic prices wages, rents and the supply of factors of production will continue to adjust to prevailing demand conditions on world markets over the longer term, while exchange rates adjustments will equilibriate the balance of payments on current and capital account. A country will be 'competitive' in the production of an internationally tradeable good or service when it can produce that good or service at a domestic price that is equivalent to, or below, the 'world market price' for the good or service in question (after allowing for quality differences) when converted from domestic prices into 'world market prices' at the prevailing exchange rate. Under these circumstances domestic producers of the good will be able to withstand competition from imports in the domestic market and to compete with other producers in export markets.

As a result, price and trade liberalisation including the exposure of the domestic economy to external competition and world market prices will necessitate a major restructuring of the CEE economies, as industrial sectors that fail to meet the conditions outlined above are forced into bankruptcy, and investment is channelled to sectors that will be profitable in the long run. This process will be accelerated by the need to comply with the requirements of the single market and EU competition policy. This implies that the economies that adopted, a path of development that was least suited to their underlying comparative advantage will experience the greatest adjustment costs following the collapse of

the CMEA market for their industrial products and the ending of state
subsidies to loss-making industries.

2.2   Factor proportions and the Heckscher–Ohlin theory of factor proportions

What will the impact of increasing cohesion with the EU on the CEE
economies? Until the late 1970s, conventional theories of international
trade were largely based on the Heckscher–Ohlin theory of factor
proportions (Ohlin, 1933). The traditional Heckscher–Ohlin theorem
asserted that trade between countries depended on differences in the
relative endowment of factors of production . A given country would be
better-suited to the production of goods and services that made intensive use of factors of production in which the country is relatively
favourably endowed. The Heckscher–Ohlin theory was based on the
assumptions of constant returns to scale and the relative immobility of
factors of production. Under these circumstances prices of goods were
determined by the price of factors used in their production, which
were in turn affected by their relative scarcity or abundance in the
domestic market. Economies that were labour-abundant, would have
a low relative price of labour and would export goods produced
by labour-intensive methods and import goods produced by capital-
intensive methods. Similarly economies that were relatively capital-
abundant would tend to export goods produced by capital-intensive
methods and import goods produced by labour-intensive methods.
Over time, however the impact of international trade will lead to identical relative prices between different goods in whichever market they are
traded which would result in the equalisation of prices of factors of
production in different countries as international trade will bid up the
demand for relatively abundant factors of production in all countries
and bid down the demand for those factors in countries in which it is
scarce.[2]

   It has been agued that the CEE economies inherited a physical capital
stock and a human capital stock (including managerial and industrial
skills) that had been explicitly created to meet a pattern of demand
which had been determined by central planners responding to political preferences and not the structure of demand determined by
consumers in a competitive, market-determined environment. This
implies that much of the physical and human capital stock inherited
by the CEE economies from communism was obsolete under the
demand conditions prevailing in western Europe.[3] Consequently the

CEE economies can be considered to be scarce in both human and physical capital in relation to the size of the industrial labour-force that they inherited from communism and would be expected to export goods produced by labour-intensive processes in exchange for good produced by processes that are intensive in human and physical capital.

According to the traditional Heckscher–Ohlin theory, trade could continue along lines which are largely determined by factor endowments as long as barriers to factor mobility remain. Under these circumstances, international trade will be 'two-way' and will involve the exchange of dissimilar goods, produced by different processes of production. However, the removal of barriers to labour and capital mobility as the CEE economies prepare for entry into the EU will result in increased movements in factors of production. Labour will be encouraged to migrate (temporarily or permanently) from labour-abundant regions to labour-scarce regions in response to the incentive of higher wage rates.[4] Financial and physical capital will flow from capital-abundant regions to capital-scarce regions in response to the lure of increased returns. The improvement in factor proportions which results from increased inflows of capital to labour-abundant, but capital-scarce regions, will contribute to a faster rate of growth of labour productivity which will, in turn, result in a higher rate of return on capital. Consequently, it is a logical inference from the Heckscher–Ohlin theory that the removal of barriers to trade and factor mobility resulting from the entry into a single, integrated market will result in a gradual but sustained reduction of price differences for products and factors of production between member states which will help to reduce differences between income levels. In the long-term the movement of factors of production will contribute to a reduction in trade that is determined by factor prices and factor intensities (which involves exports of low-technology, labour-intensive goods, possibly dependent on low relative wages from central-east and south-east Europe trade in exchange for capital and technology-intensive goods), and should eventually lead to the elimination of differences beween the prices of factors of production. An additional problem concerning the distribution of income under Heckscher–Ohlin assumptions was developed by Stolper and Samuelson (1941) who demonstrated that import liberalisation (in any given country) would bid up the demand (and price) for the abundant factor of production and bid down the demand (and price) for the scarce factor of production, to the extent that the returns to the scarce factor of production could actually be improved by protection.

2.3   The new international economics

In practice, the predictive powers of the Heckscher–Ohlin theorem appear to have been limited. Empirical studies have not found much support for the predictions of the Heckscher–Ohlin theorem that international trade flows are in fact determined by differences in domestic factor proportions between countries. At the same time substantial differences in absolute levels of wages between countries have persisted despite global trade liberalisation. Leontieff (1953), for example, discovered that the US was a net exporter of labour-intensive goods and a net importer of capital-intensive goods in the immediate post-war period. Subsequent studies of the net factor content of US trade flows in the 1960s and 1970s discovered that US exports were intensive in human capital, which cannot be considered to be 'natural' factor of production as such (Stern and Maskus, 1981). This has caused economists to return to Ricardian concepts such as relative differences between the levels of productivity in different industries in different economies to explain trade flows while the absolute level of productivity explains the persistence of international differences in the returns to factors of production. If Germany is relatively more efficient at producing sophisticated machinery in relation to steel than Romania, Germany will export sophisticated machinery to Romania and import steel from Romania, even though labour productivity is higher in both industries in Germany. Differences in wage rates between Germany and Romania will reflect the absolute level of labour productivity in both countries. Nevertheless an important implication of the theorem developed by Stolper and Samuelson remains valid under these circumstances. If German exports of sophisticated machinery to Romania embody human capital-intensive processes and are exchanged for steel which embodies labour-intensive processes, this will increase the demand for human capital-intensive labour in Germany and its wage in relation to unskilled labour and will increase the demand for unskilled labour in Romania and its wage in relation to skilled labour, compared with a situation in which no trade takes place. This tendency will be intensified by any increase in factor mobility brought about by the single market.

The question of why Germany may be relatively more efficient in the production of advanced machinery than steel than Romania remains and requires further explanation. The school of thought known as the 'new international economics' places greater emphasis on market imperfections, the process of innovation, internal and external economies of scale, product cycles and the role of consumer preferences as determinants of trade flows between countries. The most important theoretical contributions of the new international economics have been directed at explaining why trade takes place between countries at similar levels of development with similar factor proportions rather than explaining trade flows between countries at different stages of development. One problem arising from the 'new international economics' is that it does not result in a general set of predictions that can be applied to all circumstances. Although the major predictions resultimg from the analysis contained in the 'new international economics' tend to support trade liberalisation, rather than protection, the theory can result in different predictions about the nature of trade relations and income distribution which will result from trade liberalisation between countries at different stages of development than those derived from the pure Heckscher–Ohlin theorem.

In practice, a significant volume of trade between advanced industrial economies at similar levels of development consists of intra-industry trade (trade in similar products such as cars, electronic goods, household equipment) where countries simultaneously import and export the same, or similar, products. Intra-industry trade can be further subdivided into horizontally-differentiated trade (or horizontal intra-industry trade) and vertically-differentiated trade (vertical intra-industry trade). Horizontal intra-industry trade involves the simultaneous import and export of highly similar products, of similar quality levels, which are primarily differentiated by product specifications. The first attempts to explain this form of trade were made in the early 1980s by Lancaster (1980), Helpman (1980), Krugman (1981) and Greenaway and Milner (1986). The burden of explanation for this form of trade which cannot be attributed to differences in factor proportions rests on market imperfections, internal and external economies of scale and consumer tastes. These factors combine to create a small number of international producers who operate under conditions of oligopoly or monopolistic competition and whose products satisfy the specific preferences of groups of consumers located in different countries. The number of producers in any given industry will be determined by the importance of economies of scale and the size of the market which will determine the optimum scale of production. The greater the importance of economies of scale in relation to a given market, the smaller will be the number of producers who can operate at minimum average cost. Internal economies of scale can arise from high levels of fixed costs (associated with research and development, advertising, quality control, expenditure on compliance with government/EU regulations) which result in significant reductions in unit costs for large-scale manufacturers. Manufacturers who are faced with a limited domestic market for their product seek foreign markets in order to benefit from economies of scale. The incursion into foreign markets, particularly at similar levels of development, will bring the company into competition with other manufacturers of similar products. Under conditions of oligipoly or monopolistic competition, multinational corporations may choose to compete on grounds of product differentiation, catering to common consumers tastes in different countries and building up brand loyalty, rather than by reducing prices. Consequently a good may be produced in country A and be exported to consumers in country B at the same time as a similar, but brand-differentiated good with different product charactersitics, is produced in country B and is exported to country A. This is most likely to arise in trade between countries with similar income levels and similar costs of production, otherwise production would be transferred to the lower cost area of production, or the majority of consumers would substitute lower priced goods for higher priced goods of similar quality.

Although intra-industry trade in similar products may dissipate some of the benefits of economies of scale by increasing the number of producers, economies of scale may continue to arise from a combination of standardisation of production, advertising and distribution and sales to third markets which do not have major domestic competitors. The economies of scale enjoyed by existing international producers makes it difficult, but not impossible, for existing domestic producers to sustain market share when faced with international competition in the home market, or for new producers to break into domestic markets with competing products.

Although vertical intra-industry trade may appear to be relatively close to horizontal intra-industry trade, the explanation for, and implications of, this form of trade are closer to those for inter-industry trade. In vertical intra-industry trade, products are differentiated by quality factors which reflect real differences in the process of production (Falvey, 1981). These are reflected in major differences in the prices of the products which fall in the same commodity range which are imported and exported by any given country. These, may in turn, reflect differences in technological capabilities. Trade in passenger cars, for example would involve the export of higher quality cars from the capital-intensive country to high income consumers in labour-intensive countries, while the labour-intensive country would export lower quality cars to mass markets in the capital-intensive country. The higher-priced car would contain sophisticated inputs which embody a high degree of human capital while the lower-priced car may use inputs which embody labour-intensive processes. Under these circumstances the fundamental explanation for trade is the same as in the example given above where Germany exported sophisticated machinery to Romania and imported steel from Romania.

This will result in vertically-differentiated intra-industry trade which is based on differences in factor inputs and which will result in long-term wage and income differentials between countries that are intensive in human capital and those that are intensive in unskilled labour when wage rates are compared at prevailing exchange rates. However, the theory of comparative advantage indicates that this trade will result in an increase in welfare for both countries, while the Stolper–Samuelson effect indicates that relative wages in labour-intensive countries will rise relative to wages in the human capital-intensive sector, provided that labour is sufficiently mobile to respond to the change in the structure of demand for labour-intensive goods. Provided that this is the case, real wages of the lower-paid in the labour-abundant economies will rise as a result of trade.

2.3.1 Technology gaps, quality ladders and agglomeration effects in trade between countries at different levels of development

The idea that trade between countries at different levels of development will be driven by technological superiority predates the 'new international economics' but has been incorporated into some of its models. These theories have implications for the distribution of income between countries that are not present under traditional theories of comparative advantage. A country that is intensive in human capital may strive to enter a process of continuous innovation to maintain high levels of productivity and to preserve high relative incomes by concentrating on the production of innovation-intensive products. Products whose production processes have become standardised, and no longer require the use of high levels of human capital, are then produced in labour-abundant regions with low relative wages. The idea that a comparative advantage based on technological superiority could be created in a country that had a more efficient system for supporting research and development and promoting innovation was first developed by Posner (1961) and was formalised in the product-cycle model by Vernon (1966). The Posner–Vernon model provided an ad-hoc explanation of US technological superiority in the 1960s and of the international diffusion of technology by multinational corporations. The prospect of short-run supranormal profits provides the incentive for a firm to innovate and produce new high-quality products (with a high income-elasticity of demand) which are directed at relatively small, high-income markets. The initial stages of production require a highly-qualified labour force capable of meeting strict quality standards and solving technical problems related to new innovations which can command high wages. In many cases, the initial stage of production may neccesitate physical proximity to R&D facilities, requiring location in high-wage economies. The provision of capital to finance high initial levels of research and development will also require an efficient capital market, a functioning regulatory system to ensure that investors are not misled by fraudulent claims and a fiscal system that encourages investment. In the first stage of production, the product will be primarily geared to the domestic market, but may also be exported to high-income consumers in other countries. As the product becomes standardised and sales are directed at lower income markets at home and abroad, the need to reduce unit costs causes production to be transferred to regions with abundant (but technically-competent) supplies of labour and proximity to cheap sources of materials. At this stage competitors, from home and abroad, may also enter the market with similar, but differentiated products. This gives rise to the development of 'quality ladders' (Grossman and Helpman, 1991) in which a country which is intensive in human capital perpetually innovates to produce higher-quality, high-priced exports, while importing lower-quality, standardised products from labour-abundant low-wage sources. It should be remembered that such technological superiority in the methods used to produce a specific product will only contribute to the 'competitiveness' of given product if it either contributes to lower costs of production (e.g. by displacing high cost inputs) or to a higher quality product which can command higher prices. A country that is labour-abundant could remain competitive in the production of a given good by using labour-intensive methods of production that would be considered obsolete in labour-scarce countries with high relative wages (factor reversal).

The argument above does not preclude the possibility that countries may overcome an initial disadvantage and move up the quality ladder, as a result of an improved research climate, improvements to education and training and, more specifically in the case of transition economies, through changes in the economic environment away from an innovation-averse culture towards an environment that encourages innovation and improvements in the functioning of capital markets. Similarly, if the advantages of a domestic production base outlined by Posner (above) are not present, there is no *a priori* reason why a multinational corporation should not seek to establish production bases for goods which are predominantly human capital-intensive but also require some labour-intensive processes in a labour-intensive country as long as differences persist in labour costs.

Theories based on the 'new international economics' and locational theory (e.g. Porter, 1990; Krugman, 1991) indicate that first-movers in a given sector or region, can benefit from internal and external economies of scale that latecomers cannot imitate. Industrial concentrations which have developed external economies of scale which are specific to a particular location, retain a cost advantage which makes it difficult for competitors in other locations to enter the market. This enables the region that benefits from 'agglomeration effects' to survive in a pre-eminent position after the initial advantage which attracted industry to the location has disappeared. Agglomeration effects can result in the concentration of human capital in a given region. These include the provision of education and training (including the development of specialised, sector-specific training facilities), the provision of specialised infrastructure and the attraction of workers with skills, or workers who wish to obtain sector-specific skills, to the region. The 'monolocation' of human capital-intensive firms and facilities results in the continuation of inter-industry and vertical intra-industry trade based largely on exports of human capital-intensive goods in exchange for labour-intensive goods.

2.3.2 The implications of first-mover effects for EU enlargement

How will the process of enlargement and the preparations for enlargement affect the structure of trade between the CEE economies and the EU and the process of economic development in the CEE economies. Krugman (1991) has demonstrated that specific industries in the USA, where industrialisation has occured within the framework of a single market, tend to be far more concentrated in a few localities than in Europe, where separate industries developed inside the protection of national boundaries.[5] This has limited the ability of European industries, faced with relatively small domestic markets to fully benefit from economies of scale, although this has been partially offset by the development of horizontal intra-industry trade. The implications is that the progressive removal of barriers to trade and factor movments within the EU will contribute to greater industrial concentration as industrial plants seek the scale and locaton that offers optimum internal and external economies. This process will affect the CEE economies as they become integrated into the single market. Although there will be significant short-run adjustment costs as less efficient producers are forced out of business, the development of more specialised industries enjoying internal and external economies of scale within new locations will lead to an increase in aggregate economic welfare in the long term.

Will this result in a wider dispersion of capital throughout Europe, which will contribute to an increase in living standards in the poorer regions of Europe, or will it result in a process of industrial concentration in which some regions acquire economic pre-eminence at the expense of others that go into long-term economic decline? More specifically, what will be the impact of this process on economic development in the CEE economies that will be entering the EU with income levels that are substantially below those of even the poorest of the existing members? Traditional international trade theory anticipates that the removal of barriers to the movement of factors of production will result in the movement of capital from capital-intensive regions to labour-intensive regions and the movement of labour from labour-intensive regions to areas of labour shortage which will result in a narrowing of income differentials and increases in real wages in the poorest regions. Since the early 1960s, international trade theorists have developed models that argue that comparative advantage can result from productivity differentials between countries and regions which can result in enduring differences in real wages and incomes. Regions with an inherited productivity gap may experience difficulties in attracting the capital inflows which are required to raise the level of productivity in the domestic economy. This will be particularly acute when low levels of productivity result from factors that are external to the investing enterprise such as an unfavourable location which increases transport costs, poor infrastructure and communications, poor levels of general education, the effects of isolation from international markets and technology and the absence of an industrial culture which cannot be overcome without significant public investment and changes in attitudes. Locational theory suggests that regions which are the first to attract industrial concentrations can generate economies of scale that are external to the firm which cannot be replicated by regions that are latecomers. These external economies frequently relate to the provision of education and training and an economic environment that stimulates the development of human capital. This raises the possibility that the process of economic concentration will create wider income disparities and long-lasting economic divisions in Europe as capital flows are concentrated on leading regions which specialise in the production of human capital-intensive, high-wage goods while backward regions remain unsuccessful in attracting capital and concentrate on the production of lower-quality, labour-intensive goods which attract relatively low wages. Consequently economies and regions which are slower to adapt to changes in the economic environment in Europe and are less successful in attracting capital inflows will be permanently disadvantaged.

It is also possible that existing firms situated in 'latecomer' regions will be unable to sell products in the short run in sufficient volume to benefit from economies of scale, or to justify expenditure on technological and quality improvements and may be forced out of business, even though they may have a comparative advantage in the production of those goods in relation to other EU producers in the long run. The regional problem would be compounded by the free mobility of labour in the single market which could result in the migration of skilled labour from backward regions to established regions of production, lured by higher wages, or improved work prospects and opportunities.

The probability of such negative outcomes should not be overestimated. Firms and industries that are currently uncompetive in EU markets in areas in which they may have a long-term comparative advantage for reasons inherited from central-planning (e.g. poor management, over-staffing, obsolete equipment, low-quality production, absence of market recognition etc.) may become competitive following changes in owner-ship and the system of management. Many of the economies of scale enjoyed by multinational companies result from factors which are inter-nal to the firm (e.g. brand-name loyalty, research and development, managerial economies etc.) which can be separated from the process of production. Under these circumstances existing EU producers will have an incentive to relocate production to the area that offers the lowest production costs in the long-term, either by taking over existing produ-cers (brownfield investment) or through the creation of new bases of production (greenfield investment) which will create external econom-ies for other firms investing in the host coutnry which will stimulate agglomeration effects. However this could also result in a situation where human capital-intensive processes (research and development, management) are conducted in separate locations, while labour-inten-sive processes are conducted in labour-abundant regions.

Finally, there is a danger that a phased accession will exacerbate exist-ing differences in income levels and contribute to a widening of the economic gap between the states that are included and those that are excluded from the first tier of enlargement. Although the states that are not included in the first round of enlargement will benefit from inclusion in a free trade area with the EU, they will still be subject to provisions concerning 'rules of origin' and anti-dumping legislation and will face greater obstacles to exporting to the EU which may deter FDI. In addition the states that are excluded from the first round of enlargement will receive substantially lower financial transfers from the EU which exacerbate productivity differentials arising from poor infrastructure and lower expenditure on education and training.

2.4   The importance of intra-industry trade in the EU and between the EU and the CEE

How important are intra-industry trade, and horizontal intra-industry trade between the CEE economies and the EU and how does this com- pare with existing patterns of trade inside the EU? A study of trade patterns inside the single market, commissioned by the EU, indicates that in 1994 intra-industry trade accounted for between 60 and 68 per cent of intra-EU trade for the northern tier of the EU-12 (defined as France, Germany, Belgium, Luxembourg, the UK and Netherlands). Horizontal intra-industry trade ranged from between 16.5 per cent (UK) and 24.1 per cent (France) of each country's total intra-EU trade (EU–CEPII, 1997, 4–5). However, the study provides mixed evidence of the impact of participation in the single market on the trade structures of the lower-income, southern-tier economies of Portugal, Spain and Greece whose experience might be considered to offer better indications of what the CEE economies might expect following entry to the EU. Spain had a relatively high proportion of both intra-industry trade within the EU (54.1 per cent) and horizontal intra-industry trade which accounted for 18.9 per cent of total trade. Intra-industry trade accounted for only 14 per cent of trade between Greece and the rest of the EU and horizontal intra-industry trade for only 3.7 per cent. Portugal had intermediate proportions of intra-industry trade (31.4 per cent) and horizontal intra-industry trade (7.5 per cent). The proportion of hori- zontal intra-industry trade in Spain's trade with other EU economies rose by 8.7 per cent between 1987 and 1994. The study finds that Spain has benefitted from the growth of foreign direct investment (FDI) which has contributed to an upgrading of human skills and technology. This has been reflected in changes in the structure of exports to the EU towards a greater proportion of higher and medium technology goods at the expense of low technology, labour-intensive manufactured goods dependent on low relative wages (EU–CEPII, 1997, ch. 8). This provides some support for the more optimistic inferences from the Heckscher– Ohlin theorem that capital will flow from capital-intensive countries to labour-intensive countries, resulting in a qualitative improvement in the level of output. However, the increase in FDI has brought some disadvantages to the Spanish economy, including the tendency of multinational corporations to source inputs to new ventures in Spain from other EU sources at the expense of Spanish producers, while Spain has continued to experience consistently high rates of unemployment.

A major study of the extent and nature of intra-industry trade between 39 economies (including the CEE) and nine EU economies who are treated as a single unit for each of the years between 1990 and 1995 has been conducted by Aturupane, Djankov and Hoekman (1997) under the direction of the World Bank.[6] Wolfmayr-Schnitzer (1988b) has also conducted a study of intra-industry trade between inidvidual CEE economies (excluding the former Soviet Union) and the OECD as a whole, which is treated as a single unit for 1993. The methods and data sources of these two studies differ from each other, and from those used in the CEPII study and consequently the numerical results of the three studies cannot be compared directly. Nevertheless some broad conclusions can be drawn. Aturupane, Djankov and Hoekman, show that the level of intra-industry trade between the EU-9 and the CEEs was lower than that between the EU-9 and higher income economies (including Austria, Switzerland, the USA and Japan) but was higher than that in bilateral trade between the EU-9 and other lower-income non-European trade partners. Wolfmayr-Schnitzer also found that the level of intra-industry trade between the CEEs and OECD economies ranged from under a third (Romania) to around 60 per cent (Czech Republic) of the level of intra-industry trade between OECD economies, but was consistent with the levels of intra-industry trade between the OECD and non-OECD economies. Both studies also found that vertical intra-industry trade accounted for a far greater proportion of intra-industry trade between the CEE economies and the EU/OECD than in trade between the EU/OECD and comparator countries. Vertical intra-industry trade accounted for 80–90 per cent of total intra-industry trade between the EU and the CEE economies, with the latter predominantly exporting lower quality, labour-intensive products. The corollary was the proportion of horizontal intra-industry trade between the EU and the CEE economies was substantially lower than in EU trade with comparator countries. Aturupane, Djankov and Hoekman found that levels of horizotal intra-industry trade between the EU-9 and the CEE were roughly half the level of that between the EU-9 and Austria, Spain and Switzerland and were similar to levels for EU trade with Greece, India and Tunisia.

Some common features emerged between the Wolfmayr-Schnitzer study and the World Bank study on the ranking of the individual CEE economies with respect to levels of both intra-industry trade and horizontal intra-industry trade which are summarised in Table 2.1. The higher-income CEE economies (Czech Republic, Slovenia and Hungary) recorded the highest levels of intra-industry trade of the CEE economies in both studies, with the Czech Republic having the highest level in both studies. The lower income economies of Bulgaria and Romania occupied the bottom positions in both studies. The intermediate-income economies of Poland and Slovakia occupied intermediate positions in both studies, although their positions relative to one another were reversed.

Slovenia and the Czech Republic occupied first and second place in tests on horizontal intra-industry trade followed by Slovakia. The difference between the results for Bulgaria in horizontal intra-industry trade in the two studies can be attributed to an abnormally high level of horizontal intra-industry trade between Bulgaria and the EU in 1993, which was observed in both studies, but which was the only year included in the Wolfmayr-Schnitzer study. If the figures for Bulgaria in 1993 are considered to be abnormal, the remaining countries came in the order of Hungary, Romania, and Poland in both studies. Aturupane, Djankov and Hoekman also included measures for intra-industry trade and horizontal intra-industry trade for the three southern-tier economies of Spain, Portugal and Greece which confirmed the rankings of the EU–CEPII study. The study by Aturupane, Djankov and Hoekman also provides a ranking of the estimates for Spain, Portugal and Greece in

Table 2.1   Rankings of CEE economies by intra-industry and horizontal intra-industry trade

	Intra-industry trade		Horizontal intra-industry trade	
	Wolfmayr-Schnitzer	**World Bank (1995)**	**Wolfmayr-Schnitzer**	**World Bank (1995)**
Czech	1	1	2	2
Hungary	2	3	5	4
Slovenia	3	2	1	1
Slovakia	4	5	3	3
Poland	5	4	7	6
Bulgaria	6	6	4	7
Romania	7	7	6	5

Sources: Wolfmayr-Schnitzer (1998b). World Bank: Aturupane, Djankov and Hoekman (1997).

relation to the CEE economies which can help to compare the CEE economies with the rnakings in the CEPII study. Spain had levels of intra-industry trade and horizontal intra-industry trade that were sub-stantially higher than those for any CEE economy. Portugal had levels of intra-industry trade which place it behind the Czech Republic, Hungary and Slovenia, but slightly above Poland and Slovakia, while the level of intra-industry trade conducted between Greece and the EU actually fell below the level of all of the accession states. Albania provides an inter-esting paradox in the World Bank study with relatively high levels of intra-industry and horizontal intra-industry trade with the EU. This reflects the high level of trade with Greece, a relatively low-income economy by EU standards. However when horizontal intra-industry trade was measured, all CEE economies recorded lower levels than Spain and only the Czech Republic and Slovenia had had higher levels than Portugal, while Poland, Romania and Bulgaria had lower levels than Greece.

2.5   The nature of trade relations between the EU and the CEEs: empirical studies

Neven (1995) demonstrated that CEE net exports to the EU in the period 1985–90 were concentrated on products that were intensive in relatively unskilled labour and were characterised by low-wage costs. These involved processes that were either labour-intensive with a low level of investment or processes that were labour-intensive with a relatively high degree of physical capital but a low degree of human capital. CEE net imports from the EU were concentrated on goods that were intensive in human capital with high wage-rates, either combined with high levels of investment involving high technology, or in industries with a relatively low level of capitalisation. This implies that the factor content of trade relations between the CEE economies and the EU in the communist period involved the exchange of products produced by human capital-intensive processes for products produced by relatively unskilled labour, rather than differences in the intensity of physical capital. Aturupane, Djankov and Hoekman found that the volume of both horizontal and vertical intra-industry trade beween the CEE economies and the EU was highly correlated with the level of FDI and that in the case of vertical intra-industry trade the link was largely explained by labour intensity in CEE exports. This can be interpreted as indicating that the structure of CEEC exports to the EU is closely linked to FDI which (if one considers potential exports) in the case of vertical intra-industry trade is attracted by low wage costs. Ellingstad (1997) compares the process of FDI in Hungary in the early 1990s with FDI by US multinationals in Mexico where increases in productivity have not been met by compensating increases in real wages and suggests that CEE economies may simply be seen by multinational corporations as sources of competent but low-cost labour. Martin (1998) considers that the potential for the development of such a 'Maquiladora syndrome' is unlikely and argues that multinational corporations have started to distinguish between the different economies in the region and have started to incorporate the Czech Republic, Hungary and Poland into the 'international economy' through the process of trade and exchange rather than by integration into multinational systems of production, although other economies in the regions are lagging in this process. Prosi (1998) argues that the more advanced technical processes that are transferred by FDI are originally developed by multinational firms to meet the factor proportions and technological skills of more advanced economies, not those of labour-intensive economies. Consequently the availability of relatively cheap, unskilled labour is not a major attraction to multinational firms involved in medium and high technology sectors who are require technically competent sources of labour.

A number of studies have been conducted to analyse changes in the factor content of CEE trade with the EU and the quality structure of CEE exports (based on comparisons of unit values paid for imports from CEE economies and from other regions) in the period since the collapse of communism. There is general agreement that substantial quality gaps exist between CEE exports and the demands of EU markets, although there is some disagreement over whether quality gaps are closing in some cases. Brenton and Gros (1997) attribute most of the growth in CEE exports to the EU in the early years of the transition (1989–92) to a growing share of CEE products in EU markets but cannot find evidence that this has been associated with improvements in product quality. Landesmann and Burgstaller (1998) argued that CEE exports were concentrated in the lower quality segments of EU markets but indicated that the quality gap between the output of manufactured goods produced in the four central European economies (Hungary, the Czech Republic, Poland and Slovenia) and the EU, narrowed between 1989 and 1994, while the quality gap between exports from the Balkan economies (Bulgaria and Romania) and the EU had remained static or had actually widened. Carlin and Landesmann (1997) attribute the improvement in product quality to the more rapid progress made by the central-east European economies in macreconomic stabilisation, economic restructuring and improving enterprise performance. Kaminski (1998) argues that the commodity composition of Polish exports to the EU has undergone a significant improvement since 1992 which has included the expansion of exports of machinery and equipment to the EU and the emergence of new, more sophisticated product lines, with a diversification of the structure of Polish exports towards goods with a higher degree of processing.

Macroeconomic evidence indicates that the the gap between the economies included in the first and second tiers of enlargement in the Commission's initial selection has widened in the period since the collapse of communism. The real GDP of all five first tier entrants in 1998 expressed as a percentage of real GDP in 1989 exceeded that of all second-tier entrants with the exception of Slovakia. Both Romania and Bulgaria experienced two consecutive years of substantial negative growth in the mid-1990s. Four of the first-tier entrants (Hungary, the Czech Republic, Estonia and Slovenia) received the highest levels of cumulative inflows of FDI per capita between 1989 and 1997, with Poland in seventh place behind Latvia and Lithuania who have attracted FDI from Nordic states and whose per-capita levels were boosted by relatively small populations. The first tier economies occupied five of the top six places in levels of FDI per capita in 1997, with Latvia in third place (EBRD, 1998).The gap between the economic performance of the states included in the first tier of negotiations and the second tier is also reflected in their progress towards implementing economic reforms and restructuring. The EBRD quantifies the progress of reforms each year under a number of different headings to arrive at a composite index of reform. In 1998, the economies included in the first tier of negotiations plus Slovakia achieved the top six scores, followed by Latvia and Lithuania in intermediate positions, with Romania and Bulgaria occupying lagging positions (EBRD, 1998).

Conclusions and research questions

Empirical studies have shown that EU imports from CEE economies are concentrated in the lower quality segments of EU markets. This is reflected in the low levels of intra-industry trade and horizontal intra-industry trade between the CEE economies and the EU, compared with the structure of intra-EU trade. Approximately two-thirds of trade between the CEE economies and the EU consists of 'two-way' trade in dissimilar products which can be largely accounted for by differences in factor proportions. Only 4–8 per cent of trade between the CEE-10 and the EU consisted of horizontal intra-industry trade in similar products although this accounts for 15–24 per cent of trade between the richer EU economies themselves. There are signs of differentiation between the export performance of CEE economies in EU markets. The Czech Republic, Slovenia, and Hungary fall in a first group of economies with relatively high incomes, with a higher proportion of intra-industry trade (although Hungary had a relatively low level of horizontal intra-industry trade). Poland, Slovakia and the Baltic states fall into a second group with intermediate levels of intra-industry trade, while Romania and Bulgaria fall into third group with low incomes and low levels of intra-industry trade.

These differences in performance in EU markets reflect a combination of different output structures inherited from communism, the slower pace of reform and industrial restructuring in the lagging economies and the comparative failure of the Balkan economies in particular, to attract FDI. Does this reflect a temporary lag in the performance of the lagging economies, or is it symptomatic of deeper-seated problems that will prevent their full integation into Europeans economic structures? The dynamics of the single market imply that large-scale industry will become more concentrated in a smaller number of locations which can benefit from internal and external economies of scale. The locational theories outlined in this Chapter suggest that countries, or regions, that move quickly to attract invesmtent in new areas of development can benefit from 'first-mover effects' that generate external economies while those that are slow to respond to the changing dynamics of European markets may find themselves forced to concentrate on labour-intensive activities. This is now one of the major problems confronting the Balkan economies. Although it can be inferred that quality gaps between the Balkan states and the central-east European states could be closed by accelerating the pace of reform in the lagging states, it also raises questions about why this process has proved to be so difficult in the lagging states in the first place, while the impact of wars in former Yugoslavia, has effectively increased the 'economic distance' between the Balkan states and the central European states of the EU.

3
Statistical Tests and Problems of Measurement

3.1 Introduction

The trade flows between the CEE economies and the EU have been submitted to a number of statistical tests which form the basis of the analysis in Chapters 6 and 7. These are also referred to in Chapters 4 and 5. This Chapter will describe the nature and purpose of these tests and will discuss conceptual problems of measurement. It will conclude with a summary of the tests conducted on trade between the EU-15 and the combined CEE-10. The tests fall into three broad groups. The first set of tests measure the factor content and technological level of trade between the CEE-10 economies and the EU. These tests are used to examine the net factor content of trade flows between the EU-15 and the CEE-10 and to compare the technological structure and factor content of CEE exports to the EU with those of other exporters inside and outside the EU. The second set of tests identify the industrial sectors in which the CEE economies have a revealed comparative advantage, or more accurately, a revealed specialisation, in relation to existing members of the EU and to other economies outside the EU. This will establish the industrial sectors in which the CEE economies are currently competitive in trade with the EU and will help to identify sectors which could become important sources of exports in the future. These tests form the basis of Chapter 7. A third set of tests estimates the unit values of trade (values divided by quantities) between the CEE-10 and the EU in order to evaluate the quality levels of CEE exports to the EU in comparison with those of other exporters to the EU. These tests will be used to supplement the analysis in Chapters 5, 6 and 7. Each of the above tests includes comparisons between the individual CEE-10 economies to arrive at country rankings of the CEE-10 economies. They also involve comparisons with countries inside and outside the EU to place the CEE-10 economies in an international context. Finally a linkage is made between the technological level and factor content of trade to construct 'quality ladders' in trade.

3.2   Tests of the factor content and technological level of trade

Tests of the factor content of trade involve ranking and aggregating exports and imports according to the dominant factor that is normally used in the production process. These tests have been used to assess the factor-content of trade between the CEE-10 economies and the EU and to compare the export structure of the CEE-10 economies to the EU with those of other economies inside and outside the EU. An analysis is also made of the changes in the factor content of exports of individual CEE economies to the EU since the collapse of communism. A number of tests were first conducted to establish the factor-content of trade between the CEE-10 and the EU in the course of this study, ranging from relatively simple distinctions based on a high level of aggregation of products, to more sophisticated tests involving a large number of factor groupings and a lower level of product aggregation. Results of tests using a less sophisticated indicator were published in Smith (1998). The choice of tests does not alter the ordinal ranking of individual CEE economies or their status in relation to other economies, or the direction of changes in the structure of exports since the collapse of communism significantly. However, the results of the tests that will be presented in Chapter 6 are based on the methodology developed by Wolfmayr-Schnitzer (1988), on the basis of categories established by Legler and Schulmeister which involve a low level of product aggregation. Some marginal adaptations have been made to these categories.[1]

The adaptation of the Wolfmayr-Schnitzer tests divides imports and exports of manufactured goods into three principal categories; human capital-intensive, labour-intensive and resource-intensive. The human capital-intensive category has been subdivided into products which are based on production processes which utilise high, medium and low technology. The high-technology and medium technology sectors of the human capital-intensive category have then been further subdivided into labour-intensive and capital-intensive categories. Labour-intensive goods in these categories largely involve highly-skilled labour, and Wolfmayr-Schnitzer (1998, 45) notes that, within the medium-technology category, these goods are normally of higher quality than capital-intensive goods (for example specialised medical equipment as opposed to cars or chemical products). Exports and imports of manufactured goods have been allocated to these categories by Wolfmayr-Schnitzer at the three-digit (and occasionally five-digit) level of the United Nations Standard International Trade Classification (SITC). A list of SITC classifications for manufactured goods at the three-digit level and their allocation to factor intensity categories and the unit-value of EU imports from the world and the ratio of the unit values of EU imports from the CEE-10 to EU imports from the world is given in the Appendix at the end of the book.

The initial classification of factor-intensity categories was derived by Legler and Schulmeister on the basis of a detailed analysis of production processes used in Germany in the 1980s (Wolfmayr-Schnitzer, 1998, 44–5). Human capital-intensive sectors were those in which the input of qualified labour (scientists, technicians and office and managerial employees) exceeded the average by 10 per cent. This sector included most items of machinery and equipment, chemicals other than organic chemicals and fertilisers, optical, photographic and medical equipment. Labour-intensive sectors included those with a high ratio of hours worked per unit of capital and largely consisted of light consumer goods produced with relatively low technology including textiles, clothing, footwear, furniture and bedding, some sectors of transport equipment, toys and printed matter. Resource-intensive sectors included industrial products which contain a relatively high content of animal, vegetable or mineral inputs in relation to other factors and included glass, pottery, inorganic chemicals, chemical fertilisers, non-ferrous metals and iron and steel products.

It is to be expected that different processes will have been used in the production of these goods in the CEE economies, and that these categories can only be an approximation to the production processes that are actually being employed in the CEE economies. In particular, it must be expected that processes which make greater use of relatively abundant factors of production in the domestic economy will have been employed where possible (for example labour-intensive variants of processes embodying high and medium technology). Under most circumstances it will be impossible to produce products involving high technology (for example, medicines, data processing equipment) without the essential human-skills intensive component. However, it is possible that such inputs have been supplied from outside the economy (for example, through the provision of licences or the supply of components for assembly). Consequently, it is possible that these estimates underestimate the real level of trade that is dependent on differences in factor proportions.

A major advantage of the Wolfmayr-Schnitzer test is that it provides a more accurate breakdown of trade in chemical products than earlier measures of technology levels in trade flows, which tended to place all, or a significant proportion of, chemical products in the high technology sector (for example, Leamer 1984; EU–CEPII, 1997). The high proportion of chemicals in exports of manufactured goods (particularly from Bulgaria) has tended to bias estimates of CEE exports of high-technology goods upwards. However, CEE exports of chemicals to the EU contain a relatively high proportion of exports of unsophisticated chemical products, including organic chemicals and fertilisers which are classified as resource-intensive in the Wolfmayr-Schnitzer categories (see Chapter 6).

3.3   Estimates of revealed comparative advantage

Economists specialising in international trade have devised two different types of tests to estimate an economy's comparative advantage. The first type consists of indices which compare the commodity composition of exports from the given country to another country or region (for example, the EU) with the commodity composition of exports from a representative group of economies to the same country or region. These are derived from the index originally developed by Balassa (1965). Indices of this form are now more commonly referred to as export specialisation indices (ESI) as they refer only to the structure of exports and take no account of the structure of imports of the given country. The second type of tests compare the commodity composition of exports from the given country to another country or region with the commodity composition of the given country's imports from the same country or region. These are known as indices of revealed comparative advantage (RCA). These indices suffer from the disadvantage that they take no account of the relative standing of the exporting country in comparison with other exporters.[2]

Both categories of tests suffer from the disadvantage that they can only measure actual trade flows, which are affected by market distortions, as an approximation for comparative advantage. The imposition of EU quotas, tariffs or other restrictions on imports from the CEE-10, or on imports from other countries outside the EU, which distort the pattern of trade mean that the index based on actual trade flows provides a biased measurement of the relative export capabilities of the countries concerned. Similarly subsidies provided to domestic producers (either within the EU, or to exporters to the EU) which distort the pattern of EU exports and imports will result in biased measurements which do not reflect the underlying comparative advantage of exporting countries (see Chapter 7). Subsidies to EU agricultural producers arising from the CAP, and the failure to liberalise EU imports of agricultural products from the CEE-10, mean that these problems are significant in the case of trade between the CEE-10 and the EU in agricultural products. Similarly transition economies have been slower to remove subsidies for energy-producers and to close loss-making mines and power stations. These problems have been partly circumvented in the estimates of RCA indices and ESIs by confining the indices to trade in manufactured products only. Nevertheless, the impact of hidden subsidies and soft credits on CEE exports of resource-intensive manufactured goods (for example, aluminium, iron and steel goods) should not be discounted.

3.3.1 The export specialisation index (ESI)

The export specialisation index has been estimated as the ratio of the proportion of a given commodity in the exports of manufactured goods from a CEE economy (or group of CEE economies) to the EU, to the proportion of the same commodity in EU imports of manufactured goods from a given region (normally the world as a whole excluding other EU countries). In the case of trade between the CEE-10 and the EU-15, the CEE economy is considered to have an export specialisation in those industrial sectors where the sector's share in EU imports of manufactured goods from the CEE is larger than the sector's share in total EU imports of manufactured goods from outside the EU.

This is given by:

$$ESI = (xijeu/Xiteu)/(mwjeu/Mweu) \qquad (3.1)$$

where xijeu = value of country i's exports of commodity j to the EU
      Xiteu = value of country i's total exports of manufactured goods to the EU
     mwjeu = value of EU imports of commodity j from outside the EU
     Mweu = value of EU total imports of manufactured goods from outside the EU

Consequently an ESI index of greater than one in sector i indicates that the CEE economy has an export specialisation to the EU in sector i. The index shows the importance of exports from a specific industrial sector in relation to the structure of EU demand for imports for that good from outside the EU. This provides a quantitative measure of the comparative standing of CEE-10 exporters of sector i in relation to exporters from the rest of the world. A strength of the index is that it has a precise cardinal meaning. An ESI index of 2 indicates that the share of sector i in the CEE economy's exports of manufactured goods was double the share of sector i in EU imports of manufactures from the world as a whole.

The estimation of ESI indices requires a comparison between exports from the CEE and another region outside the EU and/or with other countries inside the EU. ESI indices are shown in Chapter 7 for the CEE-10 as a whole and the individual CEE-10 economies. These have been calculated in relation to total EU-15 imports from the rest of the world, excluding the EU-15, but including the CEE-10. It can be argued that the rest of the world outside the EU is not a homogeneous group as it includes other OECD economies, and non-OECD economies from three different continents and that the measurement of ESI indices in relation to total extra-EU exports to the EU does not provide a test of the relative competitiveness of EU exporters in relation to more homogeneous groups such as developed industrial economies, or newly industrialising economies. The counter-argument is that measures based on total extra-EU trade provide valuable information on the relative standing of CEE-10 exporters in relation to total EU demand for specific products from outside the EU. More critically, the omission from the comparator group of economies which are responsible for high volumes of exports of specific goods to the EU, would provide a distorted picture of the potential of CEE exporters to penetrate EU markets.

3.3.2 Revealed comparative advantage (RCA)

Indices of revealed comparative (RCA) in this Chapter have been derived by measuring the difference between the proportion of sector i in CEE exports of manufactured goods to the EU and the proportion of sector i in CEE imports of manufactured goods from the EU. This is then expressed as a fraction of the sum of the proportions of sector i in CEE exports to the EU and in CEE imports from the EU. This facilitates a comparison between product categories of different sizes and between sets and subsets of commodities.

This is given by:

$$RCA = [(xijeu/Xiteu) - (mijeu/Miteu)]/[(xijeu/Xiteu) + (mijeu/Miteu)] \quad (3.2)$$

where mijeu = imports of industrial sector j by CEE country i from the EU

Miteu = total imports of manufactured goods by country i from the EU.

The index ranges from plus one (which indicates that the CEE country does not import that good from the EU) and minus one (which indicates that the CEE country does not export that good to the EU). The critical distinction is that a positive index indicates that the CEE economy has a revealed comparative advantage in the production of the good in trade with the EU, while a negative index indicates that the CEE economy has a revealed comparative disadvantage in trade with the EU in that good.

The use of proportions of imports and exports of manufactured goods (in preference to absolute levels of export and imports of manufactured goods) is intended to eliminate, or at least to reduce, the effect of temporary imbalances in total trade in manufactured goods and to provide an indicator of a country's *relative* advantage in the production of specific manufactured goods in relation to other manufactured goods. Part of the reasoning behind this is that annual surpluses and deficits in aggregate trade flows between market economies reflect the different stages reached by individual countries in their macroeconomic cycle which will be eliminated over time by either natural corrective factors, or by changes in macroeconomic policies. Consequently the impact of aggregate imbalances on trade in individual industrial sectors (which is also reflected in the size of surpluses and deficits in trade in individual commodities) should be removed from estimates of long-term comparative advantage. At first sight, this reasoning may be harder to justify in the case of trade between the CEE-10 and the EU, in which it could be argued that EU trade surpluses reflect net capital inflows, and that EU surpluses in trade in manufactured goods could be permanent, or at least of long duration, which should be reflected in estimates of comparative advantage.

Two major counter-arguments can be put to indicate that CEE deficits in trade in manufactured goods with the EU will not be able to persist over the long term and that export and import shares should be used in preference to absolute levels of exports and imports. Firstly, EU surpluses in trade in manufactured goods were abnormally high in the mid 1990s, reaching ECU 159 billion in 1996 compared to ECU 64 billion in 1990. This indicates that the scale of EU surpluses in the mid-1990s may not be sustained in the long-run. The share of the EU surplus in trade in manufactured goods which arose from trade with the CEE-10 in 1996

(11.6 per cent) is also broadly in line with share of the CEE-10 in total EU exports of manufactured goods (10.15 per cent). Consequently, there is little reason for treating EU trade in manufactures with the CEE-10, differently from EU trade with other areas. Secondly, it can be argued that the CEE economies will be required to close, or substantially reduce, the imbalance in their trade in manufactured goods with the EU over the long term unless they can finance these deficits by running long-term surpluses in trade with non-EU members and/or by continuing to attract sufficiently large net capital inflows. The combined CEE-10 economies did run a surplus in trade with the EU-15 in trade in primary goods (including non-ferrous metal products) in 1996 of ECU 2.4 billion. This included a small surplus (ECU 35 million) in trade in agricultural and forestry products. However, a major increase in the CEE-10 surplus in trade with the EU in agricultural and forestry products appears unlikely unless major changes are made to the EU system of protecting domestic agriculture. Although the CEE-10 also recorded a surplus of ECU 1.6 billion in trade with the EU in fuel and energy in 1996, this partly reflected the re-export of refined oil products based on imported energy from Russia, which may not be long-lasting. In addition, the CEE-10 surplus in trade in primary goods with the EU was offset by deficits in services (especially financial services) which may be expected to persist over the medium term. Furthermore, the CEE-10 deficit in trade in manufactures in 1996 could not be purely associated with imports of capital goods which were financed by net capital inflows (see Chapter 5). This implies that some, or all, of the CEE economies will be required to reduce their imbalance in trade in manufactured goods over the long-term which will require them to keep the rate of growth of imports of manufactured goods from the EU below the rate of growth of exports to the EU. Under these circumstances a measure which reflects the relative shares of specific goods in exports and imports will provide a better measure of the areas of comparative advantage than a measure based on absolute levels of exports and imports. However, this does mean that the CEE economies will be credited with an RCA index of greater than unity in several industrial sectors in which trade with the EU is in substantial deficit.[3]

3.4   The application of tests of revealed comparative advantage to trade between the CEE-10 and the EU-15

Which of the ESI and the RCA indices provides the better insight into the relative competitiveness of the CEE economies? The RCA index provides a useful test for indicating the industrial sectors in which the CEE economies are competitive in relation to other producers inside the EU. However, there is some evidence (see Chapter 5) that bilateral trade imbalances between the EU-15 and the CEE-10 are more greatly influenced by supply and demand factors that are specific to the EU economies than by factors that are specific to the CEE economies. Consequently the index does not provide an accurate guide to the specific capabilities of the CEE economies in comparison with other economies and exporters outside the EU. It would be possible for the CEE economies to record high positive RCA indices in trade with the EU-15 in industrial sectors in which the EU economies do not produce the commodity in question, but in which the CEE economies are not competitive in comparison with other producers outside the EU. Under these circumstances the ESI indicator, which takes account of competition with other non-EU producers, provides a better guide to the export capabilities of the CEE economies. However, the ESI index in turn, suffers from the disadvantage that it does not take any account of the structure of CEE-10 imports from the EU. Consequently, the index does not provide any information on the relative competitiveness of CEE-10 exporters in relation to EU exporters. Consequently, in industrial sectors which are dominated by EU producers, with low levels of extra-EU production, or where trade is largely confined to the European continent (notably where transport costs are high in relation to value), it would be possible for the CEE economies to combine a high ESI index with a negative RCA index. Under these circumstances, the RCA index would provide a better indicator of the competitive pressures that the CEE economies will encounter as their economies become more closely integrated with the EU and barriers to trade are dismantled.

Table 3.1 provides a summary of estimates of ESI and RCA indices for the trade of the combined CEE-10 with the EU-15 in 1996 in 169 three-digit sectors of the SITC classification for manufactured goods. The Table shows that the two indices do not provide a clear, unambiguous identification of comparative advantage in just over 30 per cent of the industrial sectors in which trade was recorded. The CEE-10 had a clear comparative advantage according to both indices in 52 of the 169 three-digit sectors and a clear comparative disadvantage according to both indices in 65 sectors. However, the results were ambiguous for the remaining 52 categories. The CEE-10 recorded a comparative advantage according to the ESI index, but a disadvantage according to the RCA index in 32 industrial sectors. Conversely, the CEE-10 recorded a

Table 3.1   Distribution of CEE-10 comparative advantage in manufactures in 1996 by three-digit SITC categories

Category	SITC 5	SITC 6	SITC 7	SITC 8	Total
Comparative advantage to CEE	7	25	12	8	52
Comparative disadvantage to CEE	17	15	19	14	65
ESI>1 and RCA negative	3	11	14	4	32
ESI<1 and RCA positive	6	2	6	6	20
Total	33	53	51	32	169

Source: estimated from COMEXT database.

comparative advantage according to the RCA index, but a comparative disadvantage according to the ESI index, in 20 industrial sectors. In the latter case, the results were not significantly affected by the use of proportions of trade instead of absolute trade volumes to estimate the RCA index. The CEE-10 recorded a deficit in trade with the EU, which would have resulted in a negative RCA indicator if the indices had been based on export and import volumes instead of export and import shares, in only four of the 20 sectors. The failure of the two tests to provide a clear, unambiguous indication of the comparative advantage of the CEE economies in such a large proportion of industrial sectors, poses major problems of interpretation which are examined in greater detail in Chapter 7, where it will argued that trade in an industrial category must satisfy both criteria to indicate the possibility of a genuine comparative advantage.

Another problem is that the size of the ESI index does not, by its own, provide a cardinal, or even ordinal, measure of the comparative advantage of different industrial sectors for any given country, or even to a country's relative international standing in any given sector or industry.[4] Yeats (1985) demonstrates that the distribution of the index varies substantially between different industrial sectors. International trade in some industrial sectors is characterised by a cluster of countries which have an ESI index close to one, while trade in other industrial sectors is characterised by wide variations between the ESI index recorded by different countries. Consequently it is possible that a country which records a high ESI index in an industry which is characterised by a small number of countries with high positive values could still be relatively a relatively insignificant exporter (or inefficient producer) of the product in relation to other major producers of the good and could be exceedingly vulnerable to sudden changes in the global supply or demand for the good. Conversely, the most competitive country in an industry where the ESI indices are clustered around unity, could record an index of just over unity, while other relatively high ranking exporters of a given good could even record indices of below unity. As a result it is not possible to use ESI indices to rank the relative competitiveness of individual industrial sectors for any given country. However, they may be used to rank the relative importance of exports from a given industrial sector between different exporting countries.[5]

3.5 Unit values

A further set of tests concern the measurement of unit values of imports and exports. Unit values have been derived by simply dividing the value of a given sector of imports or exports expressed in thousands of ECU, by the weight of the product group expressed in metric tonnes. These are then used to estimate the relative quality of exports from different countries or regions or the relative quality of trade flows between two countries or regions. A higher unit value is considered to reflect a higher degree of quality. The reasoning behind this is twofold (Aiginger, 1998; Landesmann and Burgstaller, 1988). Firstly, it is argued that if two apparently similar goods from different sources are sold at different prices in the same market, the explanation for the difference in the prices that consumers are willing to pay for the goods must lie in differences in the consumers perception of the quality of the goods in question. In the case of goods such as shirts, the price difference may reflect superior design, cutting and stitching. In the case of goods such as washing machines a higher unit value may reflect a higher range of programming facilities. Neither of these improvements in the product specification would be reflected by a compensating increase in the weight of the product. It would also be possible to estimate differences in prices for goods directly from alternative measures of volume, (including the number of units). However, this would only enable direct comparisons to be made between highly similar products (for example short-sleeved shirts against short-sleeved shirts) to ensure that observed price differences actually reflected quality differences, and did not reflect other factors such as quantity of inputs. This is so restrictive that it would prevent any realistic comparison between the majority of products that are involved in trade. Consequently, unit values based on measurements of weight facilitate comparisons between dissimilar products. The argument is that higher quality products embody a greater proportion of factors that do not make a corresponding contribution to the weight of the product. These primarily consist of inputs of human capital, reflected in the superior programming facilities incorporated in the washing machine in the example above, or reflected in products which incorporate research and development (for example, advanced pharmaceuticals compared with aspirins) or the use of precise human skills, or other design factors for which consumers are willing to pay a premium over the price of more basic products which do not incorporate these qualities.

Aiginger (1998, 94) argues that as a country's output structure moves up the 'quality ladder' from lower quality products to higher quality products this will be reflected in an increase in the unit value of that country's aggregate exports of manufactures. Consequently, differences in the unit value of exports of manufactured goods can be taken as an approximation of the relative quality differential between products traded between different countries. There are clearly limits to this argument. For example, the existence of a few commodities which embody scarce resources (notably SITC 667, precious metals, stones and pearls) could bias aggregate unit values up significantly. Similarly a small number of heavy products with a relatively low value can bias aggregate figures downwards by a disproportionate amount. Trade in goods with low unit values would be expected to be sensitive to transport costs. Consequently, it would be perfectly rational for trade between countries involving low transport costs to contain a higher proportion of goods with low unit values than trade between distant countries. It is also apparent from the estimates of unit values in the Appendix that some human capital-intensive goods, including high-technology products in the chemicals sector have lower unit values than some labour-intensive goods including clothing.

3.6 Outward processing trade

A major conceptual problem concerns whether to include, or exclude, trade conducted under outward processing agreements from the estimates of factor-content of trade flows and the calculation of RCA and ESI indices and unit values. Outward processing trade (OPT) involves the export of intermediate goods produced inside the EU to CEE economies for further processing and re-export to the EU. Products exported by EU members for outward processing and subsequently re-imported by the EU are not subject to EU tariffs on the re-imported component and are recorded separately in EU customs data. Consequently they can be identified in trade data (see Chapter 5). CEE exports to the EU under outward processing agreements are particularly important in the clothing industry and the inclusion of imported components in export data tends to bias the RCA index upwards in this category, as CEE imports of yarns and fibres are included in a separate SITC category. Similarly, estimates of the proportion of labour-intensive exports are biased upwards for major exporters of clothing as these exports include labour-intensive imports. Unit values of exports in the clothing industry are also affected by outward processing arrangements, as the exporting country imports components (which may have a high value) and adds a labour component which is normally weightless, (or may even subtract weight in the case of cutting and trimming) from the exported good.

An ideal measure of a country's exports for all indices would only include domestic value-added, net of *all* imported materials and components (and would even exclude the contribution of imported capital goods to final production). However, this data is simply not available for the majority of international transactions. This creates major problems of comparability. For example, data are not collected for the component of CEE exports to the EU which has been imported from non-EU partners (for example, imported crude oil, feedstocks for petrochemical products, components for assembly in manufacturing industries owned by companies outside the EU). Details of trade for outward processing between existing members of the EU are not collected by customs authorities as there is no need for tariff rebates. This makes it impossible to compare trade flows, net of flows related to outward processing, between the CEE economies and the EU with that of the countries within the EU and other members of the EU. This is particularly important when comparing the structure of trade of the CEE economies with that of Spain, Portugal and Greece. Consequently estimates that exclude outward processing trade from one set of data, but include re-imported components in another, are not comparing like with like, and will bias the value of trade conducted under outward processing agreements downwards in comparison to other conventional measures of trade. Finally, it can be argued that exports from the EU to the CEE for outward processing are genuine exports which have a physical existence. They represent intermediate goods, which the CEE economy either cannot produce competitively, or cannot produce to the agreed specifications. In some cases imports for outward processing displace domestic production (for example, textile yarns) and result in reduced domestic output and employment in the same way as normal trade. Consequently, trade conducted under outward processing agreements has been included in the preliminary estimates of the factor content of trade and in RCA and

ESI indices and unit values. The impact of outward processing trade on these magnitudes is dealt with separately.

3.7 Sources of data

The major source of data used in this study is the COMEXT database published by Eurostat, the Statistical Office of the European Communities on CD-ROM. These statistics are compiled from the trade data on intra and extra-EU trade forwarded by the statistical authorities of the member states. The Tables on commodity trade in Chapter 5 and the tests below in Chapters 6 and 7 were largely based on data for 1996, which was the latest year for which detailed commodity statistics were available when these tests were first conducted. COMEXT data for 1996 are more comprehensive than data for years before 1995 as they include the trade flows of the new members (Austria, Finland and Sweden) which are significant trade partners for some or all of the CEE-10 economies, as members of the EU. Although this creates some problems of comparability with data for years before 1995 it does not appear that the use of data for 1996 introduces any major distortions from patterns for earlier years. Further tests indicate that the trade flows of the EU-15 in 1996 do not differ significantly from EU-15 trade flows in the period 1993–6 when the EU-15 ran substantial surpluses in their merchandise trade with the outside world as a whole after incurring substantial deficits from 1988 to 1992. In 1996 the surplus in the EU-15 external trade reached ECU 43.4 billion of which ECU 16.5 billion resulted from EU surpluses in trade with the CEE-10. Despite the large aggregate surplus in the trade of the EU-15 trade in 1996, the commodity structure of EU-15 trade surpluses and deficits in 1996 is consistent with that of earlier years, but with growing EU surpluses in trade in machinery and equipment and chemicals between 1990 and 1996, part of which is explained by trade with the CEE-10.

3.8 A summary of the tests on trade between the EU-15 and the combined CEE-10

3.8.1 Factor proportions, revealed comparative advantage and export specialisation

Table 3.2 gives a summary of the basic factor proportions involved in total trade (including primary goods) between the EU-15 and the combined CEE-10 in 1996. Trade in manufactured goods has been broken

Table 3.2 Commodity and factor proportions in trade of EU-15 with CEE-10 (per cent of total)

	EU trade with CEE-10			CEE-10		
	EU Exports	EU Imports	balance	Extra-EU Imports	RCA	ESI
Primary goods						
(0,1,2,3,4)	10.76	16.46	−5.70	28.11	+0.21	0.585
agriculture & forestry						
(0,1, 21–4, 29, 4)	7.39	9.25	−1.86	10.99	+0.11	0.841
food & beverages (0,1)	5.88	5.38	+0.50	7.89	−0.04	0.682
minerals (excluding						
non-ferrous metals)						
(27,28)	0.47	1.56	−1.09	2.10	+0.54	0.740
fuel and energy (3)	2.21	5.17	−2.96	13.74	+0.40	0.377
Manufactures						
(5,6,7,8)	88.92	82.73	+6.19	69.29	−0.04	1.194
Human Capital-						
Intensive	51.44	31.34	+20.10	40.46	−0.25	0.775
High technology	8.23	3.87	+4.36	15.76	−0.36	0.246
labour-intensive	4.11	2.99	+1.12	10.04	−0.16	0.297
capital-intensive	4.13	0.88	+3.25	5.71	−0.65	0.155
Medium technology	35.88	20.40	+15.48	20.18	−0.28	1.011
labour-intensive	20.99	11.41	+9.58	13.40	−0.30	0.851
capital-intensive	14.89	8.99	+6.00	6.78	−0.25	1.326
Other	7.32	7.07	+0.25	4.52	−0.02	1.564
Labour-Intensive	27.17	33.77	−6.60	20.41	+0.11	1.654
Resource-Intensive	10.27	17.63	−7.38	8.32	+0.26	2.117
TOTAL (0–9)	100.00	100.00	0.00	100.00		

Source: estimated from COMEXT database.

down according to the Wolfmayr-Schnitzer categories outlined above. The structure of EU-15 imports from outside the EU is also shown for purposes of comparison. Column 5 consists of indices for revealed comparative advantage (RCA) estimated on the basis of factor proportions for trade between the combined CEE-10 and the EU-15 with the RCA estimated and presented from the perspective of the CEE-10. A positive RCA index indicates that the combined CEE-10 had a revealed comparative advantage in bilateral trade with the EU-15. Column 6 consists of ESI indices for the combined CEE-10 which compare the ratio of the proportion of that category in combined CEE-10 exports to the EU-15 to the proportion of that category in the imports of the EU-15 from outside the EU.

Although primary goods constituted a greater proportion of EU-15 imports from the CEE-10 (17.42 per cent) than EU exports to the CEE-10 (9.56 per cent), the proportion of primary goods in EU imports from the CEE-10 was lower than that from the rest of the world (28.11 per cent). As a result the CEE-10 economies recorded a positive RCA index in trade in primary goods with the EU-15, but an ESI index of less than one. This indicates that despite the CEE-10 surplus in trade in primary goods with the EU-15, the CEE economies had a comparative disadvantage in trade in primary goods with the EU when compared with the rest of the world outside the EU. This was also the case for all sub-sectors of primary goods, except food and beverages, where the CEE economies also incurred both a negative RCA index in trade with the EU as well as an ESI index of less than one, despite their high levels of agrarian employment in comparison with the EU.

The arithmetic counterpart of the results for trade in primary goods is that the combined CEE-10 recorded a comparative disadvantage (RCA of −0.04) in trade in manufactured goods with the EU-15, but an ESI index of greater than one (1.194). This indicates that the CEE-10 have a small comparative disadvantage in trade in manufactured goods in general in their trade with the EU-15, but that manufactured goods occupy a greater share of EU imports from the CEE-10 than from the world as whole. However, the pattern of trade in manufactured goods reveals critical weaknesses in the composition of CEE exports to the EU, both in comparison with CEE imports from the EU and with EU imports from the rest of the world. The combined CEE-10 economies had both a strong revealed comparative disadvantage in trade in human capital-intensive goods with the EU-15 as a whole (reflected in an RCA index of −0.24) and a lack of export specialisation in human capital-intensive goods in comparison with other exporters from outside the EU (reflected in an ESI of 0.775). Within the category of human capital-intensive products, the RCA (−0.361) and ESI indices (0.246) are least favourable to the CEE economies in goods that embody high-technology. Although the CEE-10 have negative RCA indices in trade in medium technology products (−0.28) and low technology goods (−0.02), the CEE-10 display a positive ESI index in trade in human capital-intensive goods which embody medium (1.01) and low-technology (1.56). Finally, the combined CEE-10 record a strong revealed comparative advantage in trade in labour-intensive goods (which primarily consist of light consumer goods) and resource-intensive goods (which largely consist of iron and steel products, non-ferrous metals and unsophisticated chemicals) according to both indicators.

3.8.2   Unit values and the construction of quality ladders

Table 3.3 provides estimates of the unit values of trade between the EU-15 and the CEE-10 in manufactured goods which have been broken down by factor proportions according to the Wolfmayr-Schnitzer definitions. Unit values for intra-EU trade and trade between the EU-15 and the rest of the world are also provided for purposes of comparison. The final two columns in Table 3.3 show the ratio of unit values in total extra-EU imports and intra-EU trade to the unit value of trade between the CEE-10 and the EU. The estimates also reveal some methodological problems resulting from the inclusion of a number of heavy items with low unit values in the category of human capital-intensive goods in the Wolfmayr-Schnitzer classifications. These have a major impact on the unit values for human capital-intensive goods as a whole, and on medium technology products in particular. These items largely consist of relatively low-value chemical products including certain inorganic chemicals, plastics in primary form and lubricating oils and preparations, together with some items of metal pipes. These have been

Table 3.3   Unit values in trade in manufactured goods in 1996 (ECUs per metric tonne)

	EU imports from			Ratio of unit values	
	CEE-10	Extra-EU	Intra-EU	Extra-EU: CEE-10	Intra-EU: CEE-10
Manufactures (5,6,7,8)	948	2,873	2,407	3.030	2.539
Human Capital-Intensive – total	1,458	5,046	4,532	3.461	3.108
without resource-intensive	2,139	8,012	9,514	3.745	4.447
High technology	2,089	19,074	17,113	9.131	8.192
labour-intensive	1,866	16,000	29,396	8.573	15.752
capital-intensive	3,502	28,794	11,836	8.222	3.380
Medium technology	1,575	4,283	3,456	2.719	2.194
without resource-intensive	3,849	12,830	9,399	3.333	2.442
labour-intensive	3,207	14,101	11,686	4.397	3.644
capital-intensive	957	1,802	2,192	1.882	2.291
(resource-intensive)	447	1,067	1,035	2.385	2.359
Low technology	1,056	1,824	5,203	1.726	4.926
Labour-Intensive	3,953	7,636	5,390	1.931	1.363
Resource-Intensive	308	621	0,644	2.017	2.091

Source: estimated from COMEXT database.

classified as resource-intensive within the human capital-intensive sector embodying medium technology. The effect is to bias the aggregate unit value of human capital-intensive goods embodying medium technology downwards. When these items are removed from the estimates of unit-values for medium technology goods and total human capital-intensive goods, a clear hierarchy of unit values can be established which forms the basis of the quality ladder set out in Table 3.4 which accords with the Wolfmayr-Schnitzer factor classifications.

It is noticeable that the unit value of EU imports of labour-intensive goods is greater than the unit value of imports of human capital-intensive goods which embody low technology, although the difference is less marked in the case of intra-EU trade. This reflects the high level of outward processing in this sector. The unit value of resource-intensive goods falling in the human capital-intensive goods embodying medium technology classification and resource-intensive *per se* are at the bottom of the ladder. This is not particularly surprising in view of the fact that resource-intensive products contain a relatively high proportion of heavy products. However, it suggests that comparisons of aggregates of unit values that contain high proportions of resource-intensive goods and/or labour-intensive goods should be treated with caution.

The outstanding feature of the unit values in Table 3.3 is the major difference between the unit value of EU imports from the CEE-10 and those for the same factor classification in both intra-EU trade and EU imports from the rest of the world. It is also noticeable that this difference is greatest in the case of high technology products where the unit values in intra-EU trade and imports from the rest of the world are 8.192 times and 9.131 times greater than the unit values of EU imports from the CEE-10. With very few exceptions, the unit values of imports of

Table 3.4   Quality ladders defined by unit values in intra-EU trade and EU trade with rest of the world

Factor category

Human capital-intensive – high technology
Human capital-intensive – medium technology – capital-intensive excluding resource-intensive
Human capital-intensive – medium technology – labour-intensive
Labour-intensive
Human capital-intensive – low technology
Human capital-intensive – medium technology – resource-intensive
Resource-intensive

individual high-technology goods in SITC three-digit categories from each of the CEE-10 economies are lower than the average unit value for EU imports from the rest of the world indicating a generalised problem in producing high-quality goods that embody high technology. The scale of the problem can be seen from the fact that the unit values of EU imports of high-technology products from the CEE-10 were actually below those for imports of medium technology goods from the CEE-10.

The gap between unit values for EU imports of labour-intensive goods and resource-intensive goods from the CEE-10 and the unit values in intra-EU trade and EU imports from the rest of the world is substantially smaller than for human capital-intensive goods. The unit values for intra-EU trade in labour-intensive goods is only 36 per cent higher than the unit value of imports from the CEE-10. Although this indicates that quality standards of CEE-10 exports (including those conducted under outward processing arrangements) are close to the average for EU production, it also suggests that EU producers may encounter problems with price competition if wage rates move towards EU levels without compensating improvements in productivity.

4
East European Economic Relations under Communism

4.1 Introduction

The foreign economic relations of the CEE-10 economies in the period between the end of the second world war and the collapse of communism were determined by the incorporation of the east European economies into the Soviet bloc and the adoption of the Soviet model of development. Following the communist takeover of power in eastern Europe in 1948 and the slide into cold war, the east European economies were forced to implement the Stalinist model of heavy industrialisation within their domestic economies. They were also required to redirect their external economic relations towards other socialist states and to the Soviet Union, in particular, and to minimise their contacts with, and dependence on, western market economies. This was accompanied by the adoption of the Soviet system of planning and its associated institutions which included a centralised state monopoly for the administration of foreign trade. This Chapter will examine the impact of these factors on East European trade relations in the communist era and the problems this has created for the redirection of trade flows to the industrialised market economies and the EU after the collapse of communism.

4.2 The Soviet growth strategy and industrial and trade priorities

4.2.1 The Soviet strategy of rapid industrialisation

The strategy of rapid industrialisation had been implemented by Stalin in the Soviet Union in the 1930s, by the creation of a centralised system of planning which subjugated market forces to central controls. The system was developed to cope with the problems created by the relative economic backwardness of the Soviet Union and to preserve Soviet power against actual and perceived internal and external threats to the security of the Soviet state. The Stalinist system generated a rapid growth of inputs of capital, labour and raw materials into industrial production by restraining private consumption and by devoting a relatively high proportion of GDP to investment. Priority was given to investment in the metallurgical, power and electrical engineering and machine tool industries and to the extraction and transportation of energy and raw materials for industrial consumption. Investment in light industrial goods and consumer goods received low priority. Investment priority was extended to include the chemicals, armaments and nuclear industries in the period following the second world war. The growth of the industrial labour force was achieved by a major expansion of female participation in industrial employment in the 1930s and by the forced collectivisation of agriculture which accelerated the exodus of labour from the countryside to the towns and newly-constructed factories. A high rate of investment was maintained throughout the Soviet period by means of forced savings. This was achieved by state controls over prices and wages and the supply of consumer goods. Forced collectivisation and compulsory deliveries of agricultural produce to the state also enabled the state to depress the prices of agricultural products below the levels that would have prevailed under market conditions.

The Soviet industrialisation strategy also made extensive use of the acquisition and diffusion of foreign technology (technical progress extended). Foreign technology was acquired both through the purchase of machinery and equipment which was used in newly-constructed enterprises (embodied technology) and through the acquisition and replication of existing world technology. The latter was achieved by both legal and illegal methods, including the employment of engineers and managers who possessed technical 'know-how', the study of technical journals and 'reverse engineering' (the acquisition of single items of machinery for the specific purpose of imitation).

This pattern of industrialisation had a number of implications for the organisation and structure of Soviet trade. Firstly, it increased the demand for physical capital and skilled labour, which were scarce factors of production in the Soviet Union in the 1930s, and consequently increased the demand for imports of goods embodying these factors. Secondly rapid industrialisation increased the demand for unskilled labour and natural resources in which the Soviet Union was relatively abundant. Unskilled labour was directed towards labour-intensive projects specified in the central plan (for example, the use of manual labour for strenuous repetitive functions in large-scale construction projects) until the capital stock had grown to provide labour with sufficient machinery to mechanise these functions. This prevented the Soviet Union from developing exports of labour-intensive goods, despite the abundance of labour, and forced the Soviet Union into greater dependence on exports of resource-intensive goods (which in the 1930s largely consisted of grain, oil and timber) to finance imports of capital. The continued emphasis on industrialisation in the 1940s and 1950s, and the neglect of agriculture, meant that by the early 1960s the Soviet Union had become a net importer of grain and foodstuffs. The growing demand for imported capital goods and foodstuffs placed additional strain on the Soviet balance of payments in the 1960s and 1970s which increased the demand for exports of resource-intensive products to maintain external balance. This necessitated further investment in the exploration, exploitation and transportation of oil and gas which, in turn, increased the demand for imports of capital equipment and pipelines for these sectors. Although, the maintenance of external balance was facilitated by the two major increases in world oil prices in 1973 and 1979, the collapse in world oil prices in 1986 resulted in growing trade deficits and Soviet indebtedness in the late 1980s which contributed to the collapse of the Soviet system. Although the Soviet Union became a major exporter of armaments and equipment for construction projects to third world countries during the 1970s and 1980s, a major proportion of Soviet armaments exports were delivered on extended credit terms or were paid for in soft currencies, if at all, and the profitability and contribution of these exports to the Soviet balance of payments is subject to considerable doubt (Smith, 1993).

4.2.2   The organisation of trade relations under central planning

The Soviet development strategy required the state to subordinate foreign economic relations to the requirements of the central plan. Imports were concentrated on goods that were of prime importance to the implementation of central plans but were difficult to produce domestically. Planners also prevented the export of goods which the state wanted to allocate to centrally-determined objectives. This was achieved by the creation of a centralised state monopoly of foreign trade which was administered by the ministry of foreign trade. The ministry of foreign trade formulated and implemented foreign trade plans, which consisted of schedules of imports and exports, in direct consultation with the state planning commission and the political authorities. The ministry of foreign trade also controlled all flows of goods into and out of the Soviet Union and effectively isolated domestic enterprises from capitalist markets and world market prices.

Prices on centrally-controlled internal Soviet markets were determined by central authorities and were entirely separated from world market prices. The domestic currency (the rouble) was inconvertible and could not be used as a means of payment for external transactions on either current, or capital, account. Consequently trade with market economies had to be conducted in convertible currencies or, in some cases, on the basis of special bilateral clearing arrangements. External financial settlements relating to foreign trade activities were conducted by a specialised foreign trade bank which was subordinated to the central state bank while the ministry of foreign trade attempted to maintain a rough balance in its trade in convertible currencies. The value of imports and exports conducted in convertible currencies were then converted into roubles at the official exchange rate, for internal accountancy purposes and for the publication of foreign trade statistics. The ministry of foreign trade conducted a separate set of operations inside the Soviet economy. It bought exports from domestic enterprises, paying for these in roubles at internal prices, and sold imported goods to domestic enterprises at internal Soviet prices. Imports normally consisted of goods that had a high domestic price in relation to world market prices while exports normally consisted of items with a low domestic price in relation to world market prices. The profits realised in domestic prices from foreign trade activities were a major source of budget revenue.

Enterprises, individuals and other government agencies of both a productive and non-productive nature (for example, administrative departments, hospitals, schools, providers of infrastructure) were not permitted to initiate any form of foreign economic activity, or to engage in foreign economic activity that had not been approved by central authorities. Consequently enterprises had very limited contacts with either foreign suppliers, or foreign markets, and acquired little knowledge about the specific demands of foreign markets including such matters as quality standards, product specifications, changes in consumer tastes, advertising, marketing and packaging. The state monopoly of foreign trade provided domestic enterprises with a safe internal market for their products which was reinforced by the absence of domestic competition. Consequently, enterprises operated in an environment in which they had no incentive to conduct research and development, to innovate either in the form of producing new or improved products, to develop improved or cost-reducing production processes, or even to imitate developments in foreign markets in order to withstand the threat of entry from domestic or foreign competitors. As a result enterprises tended to be both 'innovation averse' and 'export averse', and preferred to meet the needs of a relatively undemanding domestic market rather than to get involved in the more complicated and costly activities that were required to compete in foreign markets. This reinforced Soviet dependence on resource-based exports with a relatively low degree of processing.

Centralised controls over imports and exports were supported by strict controls over inward and outward capital flows. The absence of domestic stock markets automatically prevented inflows of portfolio investment. Domestic economic agents and individuals were not allowed to borrow from, or lend to, foreign capital markets and could only hold bank accounts denominated in foreign currencies under strictly specified circumstances. The major exception to this rule was that the state accepted credits from foreign governments (and later from the IMF and the World Bank) and from foreign banks to finance investment and current account deficits. Foreign direct investment was not permitted in east European economies until the early 1970s when western companies were allowed to establish joint ventures with minority equity participation under highly restricted circumstances.

4.3   The extension of the Soviet economic system to the Baltic states and eastern Europe

Following the outbreak of the Korean War, the central and south-east European states were required to accelerate investment in heavy engineering products (including metalworking equipment, power and electrical engineering products and equipment, transport equipment, equipment for the development of the energy sector, iron and steel, and non-ferrous metallurgical products and chemicals) and to export these products to the Soviet Union to strengthen Soviet military capabilities. Soviet economic priorities were enforced in the Baltic states by incorporating their economies into the Soviet internal planning system, with the effect that decisions concerning the structure of investment and distribution of inputs and outputs between the Baltic states and the remainder of the Soviet Union were taken by planners located in Moscow. Trade flows between the Baltic states and the remainder of the Soviet Union were not monetised but were recorded for planning purposes and enterprise accounts in Soviet internal prices denominated in roubles. Although the Baltic states were allowed to create their own economic institutions which replicated central Soviet institutions, these were directly subordinated to the Soviet institutions and had little or no autonomy. However, individual Soviet republics were not allowed to create their own ministries of foreign trade and trade flows between the Baltic states and non-Soviet states were administered centrally by the Soviet ministry of foreign trade. As balance of payments accounts for trade between Soviet republics were not prepared, there was no need for formal capital flows to offset deficits or surpluses on current account transactions between Soviet republics. Capital flows between the Baltic states and other Soviet republics simply reflected central planners decisions concerning the size and location of investment projects and the sourcing of physical inputs to those projects. Consequently the Baltic states became net recipients of inputs of physical capital which were used to build up heavy industry largely for the re-export of finished goods to other Soviet republics.

4.4   The development of trade patterns between the CMEA economies

The implementation of the Stalinist system and Stalinist industrial priorities was far more complicated in the central and south-east European states that became members of the Council for Mutual Economic Assistance (CMEA).[1] Unlike the Soviet republics, the CMEA states retained their identities as separated independent states, but were required to create the economic institutions of a centrally-planned economy on a national basis, in imitation of the Soviet planning system. Each of the CMEA economies retained its own central bank and national currency and established its own national planning agency and system of internal distribution with national sets of wholesale and retail prices. The CMEA states also created separate systems for the administration of foreign trade which involved the creation of individual national state monopolies of foreign trade and individual exchange rates linking the domestic currency to external currencies. This greatly complicated the process of determining and administering rational trade flows between the CMEA economies themselves. The foreign trade monopoly had been created in the Soviet Union to serve the needs of a single planning bureaucracy in a large country that had an abundant supply of natural resources, and was not highly dependent on foreign trade, and which wished to limit its dependence on foreign states. This system was badly-suited to the needs of small states that had been highly dependent on foreign trade in the inter-war period. Furthermore, there was no clear basis for determining trade flows between socialist states with trade monopolies. The economic logic of central planning implied that there should be a supranational body which would be capable of drawing up and implementing plans at the level of the CMEA and which would determine trade flows between the CMEA states in the same way that the Soviet central planning agency determined flows of goods and services between Soviet republics. Economic logic also indicated that there should be a single agency to plan the external trade relations of the CMEA. Despite attempts by Khrushchev to give the CMEA such supranational powers in 1962, the CMEA never became a genuine supranational agency and functioned largely as body which established the rules for trade cooperation and intra-CMEA pricing and which stimulated cooperation in joint investment projects, particularly in the energy sector.

However, the sheer size of the Soviet economy and east European dependence on Soviet supplies of energy and raw materials, combined with the political power of the Soviet Union meant that East European economic agencies were subordinated to decisions taken by their Soviet counterparts (*de facto*, if not *de jure*), in matters concerning the general direction of their patterns of production and their external economic relations. The increased production of heavy industrial goods greatly increased east European demand for energy and raw materials which could not be met from domestic sources. This resulted in the development of the 'radial' pattern of trade, whereby trade between the CMEA economies resembled a series of spokes linking the individual CMEA economies to the Soviet economy. The Soviet Union supplied its CMEA partners with energy and raw materials in exchange for imports of machinery and equipment, foodstuffs and industrial consumer goods (see below).

The development of trade links and capital flows between the non-Soviet CMEA economies was hampered by the failure to establish price systems that reflected costs of production and scarcities. This was compounded by the failure to create a properly functioning monetary system within the CMEA which could be used to settle imbalances in payments and which would have provided countries with an incentive to export. A system of prices for intra-CMEA trade was developed which was used primarily for accountancy purposes. However, the system did have an impact on the distribution of resources between countries as it affected the terms of trade between partner countries. In principle, prices for intra-CMEA trade were based on a variant of world-market prices which were converted into a unit of account which was known as the 'transferable rouble' (TR) and which was nominally equivalent in value to the Soviet foreign trade, or valuta, rouble. From the late 1950s until 1975, intra-CMEA prices were fixed in advance for the duration of each five-year plan period on the basis of an average of world market prices that had been estimated for the preceding five-year plan period that were also supposed to have been 'cleansed of monopoly and speculative factors'. The formula was changed in 1975 'for certain products' (principally oil) to an average of the preceding three years world market prices to reflect the quadrupling of world oil prices in 1973–4. The formula for calculating intra-CMEA energy prices was again changed in 1976 to a sliding average of the preceding five years world market prices. In practice actual prices charged for goods were the subject of bilateral negotiations between the respective foreign trade ministries and differed considerably from these principles. The strict application of the formula would have meant that obsolete manufactured goods which were virtually unsaleable outside the CMEA (particularly machinery and equipment) were traded at relative prices that greatly exceeded their real value, while intra-CMEA prices did not reflect the relative scarcity of raw materials and agricultural products for which an external market could be found. This lead to the practice of dividing trade between 'hard goods' which could be sold on world markets, and 'soft goods' for which there was a surplus inside the CMEA. Trade negotiators would only offer hard goods in exchange for hard goods and would only accept soft goods in exchange for the acceptance of soft goods by the trade partner. The principal exception to this practice was the Soviet Union which provided 'hard' goods (energy and raw materials) in exchange for soft goods (machinery and equipment).

The failure to create an adequate monetary system for clearing payments deficits between trade partners meant that total exports and total imports between any two trade partners were normally balanced in any one year. Settlement of imbalances in trade was normally effected by an offsetting equivalent deficit/surplus in visible trade between the two countries in a later year. The Soviet Union, which continued to run large trade surpluses with its CMEA partners throughout the 1970s and early 1980s, following the increase in intra-CMEA oil prices was also a notable exception to this rule. A number of ad hoc changes were introduced to streamline the working of the system in the 1970s and 1980s, including the use of settlements in hard currency at prices closer to current world market prices for goods that could be exported outside the CMEA. This was intended to give exporters an incentive to meet world market quality specifications for goods exported to other CMEA partners. However, enterprises which still had a guaranteed domestic market for existing produce had little incentive to meet the additional quality specifications demanded by world markets. Attempts were also made to introduce 'direct links' between enterprises in different countries in an attempt to decentralise the day-to-day conduct of foreign trade operations including delivery schedules, specification of quality standards etc. Despite these attempts, enterprises in one CMEA country frequently had very little contact with its suppliers or markets in another CMEA country and were frequently unaware of, and unresponsive to, changes in demand patterns.

As a result, the pattern of trade between CMEA economies did not necessarily reflect the comparative advantages of the individual countries. The development of more rational trade links between the CMEA economies was also hampered by the reluctance of the more agrarian economies (particularly Romania) to limit their prospects for industrialisation and concentrate more resources on the development of agricultural products for bloc consumption. Romania was not alone in taking a quasi-autarkic attitude. The more industrialised economies were also reluctant to give up lines of production which involved labour-intensive processes which would have created changes in demand for labour and would have created frictional unemployment. Central controls over foreign trade meant that they did not have to fear competition from other economies inside or outside the CMEA which had a comparative advantage in labour-intensive processes. These problems also hampered the development of genuine multinational enterprises within the CMEA and hindered the transfer and diffusion of technology between CMEA partners.

4.5   The size of intra-CMEA trade at the end of the communist era and problems of measurement

The collapse of trading arrangements that were determined by Soviet industrial priorities and the final dissolution of the CMEA in 1991 required the east European economies to seek new forms of foreign trade, the most important of which was the re-establishment of trade links with western Europe. The measurement of the importance of intra-CMEA trade for the central and south-east European economies and the estimation of the size of the task of replacing intra-CMEA trade by new trade relations is complicated by the lack of correspondence between prices used in internal (domestic) trade, in intra-CMEA trade and in trade with non-CMEA countries and by the use of different exchange rates to evaluate intra-CMEA trade and trade with world markets. This makes it impossible to make direct comparisons between the values of trade flows in different markets which operate with different price systems and to compare the value of imports and exports with domestic output during the Soviet period.

The first problem arises from the use of different prices to record intra-CMEA trade and trade with world markets in official statistics. This resulted in different *relative* prices between goods in intra-CMEA trade and trade in world markets. Consequently two bundles of goods which had an identical value in transferable roubles could have completely different values expressed in world marker prices. In the period before the collapse of world oil prices in 1986 the price ratio of most manufactured goods to those for energy and raw materials in intra-CMEA trade was substantially higher than the price ratio on world markets. The relative overvaluation of manufactured goods in intra-CMEA trade was further complicated by the fact that the majority of manufactured goods traded in CMEA markets did not meet the quality specifications demanded on world markets and were sold in EU markets at prices below those received by other exporters to the EU. However this problem had been reduced, but not entirely eliminated, by the end of the 1980s when prices for crude oil and refined oil products fell more rapidly in world markets than in intra-CMEA trade. As a result the price differential for crude oil exports to CMEA and world markets fell from roughly 100 per cent in 1988 to 40 per cent in 1989.

The second problem arose from the use of an official arbitrary exchange rate to evaluate and compare intra-CMEA trade with trade in convertible currencies. The International Bank for Economic Cooperation (IBEC) which was the official CMEA bank responsible for clearing intra-CMEA settlements, established an official exchange rate between the CMEA unit of account, the transferable rouble (TR), and a basket of western currencies. This resulted in an official exchange rate for 1988 of TR1.00:$1.646 which was used for measuring the values of intra-CMEA trade and trade with the outside world in Soviet and CMEA trade data. Trade conducted in convertible currencies was simply re-estimated at the official exchange rate to arrive at a value that could be expressed in transferable roubles. However, when intra-CMEA trade was re-evaluated at the official exchange rate it was substantially overvalued in relation to trade in world markets. By the end of the 1980s the problem of overvaluation had become so serious that Hungary, Poland and Czechoslovakia introduced separate exchange rates relating their own currencies to the transferable rouble and to outside currencies. However, the cross-rates between the transferable rouble and the dollar which resulted from this process varied substantially from country to country. The scale of the problem was such that a more than threefold difference occurred in the measurement of the dollar value of intra-CMEA trade flows between estimates based on official Soviet and CMEA exchange rates between the rouble and the dollar and estimates based on rouble-dollar cross-rates used by the Hungarian and Polish authorities. The official exchange rate of the Hungarian forint against the dollar was given as 50.413 forints to the US$ and as 26.0 forints to the transferable rouble in 1988, giving a rouble-dollar cross-rate of TR1.00:$0.52 compared with the CMEA official rate of TR1.00:$1.646. On this basis Soviet imports from Hungary in 1988 which amounted to TR 4943 million in 1988 in Soviet trade statistics would be evaluated at $8136 million according to the Soviet/CMEA official exchange rate while the same volume of Hungarian exports to the Soviet Union, which are recorded as 13,159 million forints in Hungarian trade statistics were equivalent to only $2760 million at the Hungarian exchange rate of the forint to the dollar. The corresponding figures for Soviet exports to Hungary in 1988 are $7381 million from estimates based on Soviet data and exchange rates and $2347 million from estimates from Hungarian data and exchange rates.

A similar differential occurs between Polish trade data and Soviet trade data. The official exchange rates of the Polish zloty in foreign trade statistics in 1988 is given as one US dollar being equivalent to 430.63 Polish zloty and the transferable rouble as equivalent to 97.213 Polish zloty. This give a cross-rate of one transferable rouble being worth only $0.458 which is lower than the cross-rate derived from Hungarian data. Soviet imports from Poland of 7.1 billion roubles in 1988 are equivalent to $11,700 million according to the Soviet/CMEA official exchange rate while the same volume of Polish exports to the Soviet Union in 1988 of 1475 billion zloty would be evaluated at only $3424 million according to the Polish official exchange rate of the zloty to the dollar. On the same basis, Polish imports from the Soviet Union in 1988 were worth $2853 million at the Polish exchange rate while Soviet exports to Poland are evaluated at $10,366 million. As a result Polish exports to the CMEA in 1988 are given as 72.7 per cent of total Polish exports in CMEA publications, but as only 41.1 per cent of total Polish exports in official Polish statistics.

Which of these rates provides the most accurate measure of the importance of intra-CMEA trade? The problems relating to both absolute prices and relative prices can be illustrated by comparing the absolute and relative value of Soviet exports of crude oil in Soviet trade data. The average unit value of Soviet exports of crude oil to western Europe (excluding the Federal Republic of Germany where the price was substantially higher) in 1988, was TR60 to TR63 per tonne, equivalent to $14–15 per barrel at the official Soviet and CMEA exchange rate. This roughly reflected the prevailing world market price in 1988. However, the average unit value of Soviet crude oil exports to CMEA partners was TR120 to TR125 roubles per tonne, equivalent to $28–9 per barrel, at the official CMEA exchange rate. This indicates that the transferable rouble was overvalued in intra-CMEA trade in crude oil.

It is far harder to establish clear differences between intra-CMEA prices and world market prices for manufactured goods which constituted the majority of east European exports to the Soviet Union. Firstly, many items of machinery and equipment were only traded within the CMEA which prevents a comparison between intra-CMEA prices and world market prices. Secondly, where apparently similar items were traded in both markets, the product nomenclature in official statistics (particularly for machinery and equipment, chemical products and many consumer goods) is not sufficiently detailed to permit a meaningful comparison between unit values of goods traded inside and outside the CMEA. This is reflected by major differences in the unit value of goods with the same product nomenclature (for example, metal cutting machine tools) in trade with different market economies. This probably reflects the inclusion of highly differentiated products in the same category. It is however noticeable that for some consumer goods where unit values can be estimated and which were relatively important exports from eastern Europe to the Soviet Union (for example, footwear) the Soviet import price from CMEA partner in roubles was *below* the Soviet import price of the equivalent product from western Europe and other non-CMEA partners. For example, the average price of a pair of shoes imported by the USSR in 1988 according to Soviet trade statistics was 15.35 roubles a pair from Yugoslavia (whose trade with CMEA partners was not conducted under CMEA pricing rules) and 14.93 roubles a pair from Austria compared with 10.04 roubles a pair from Poland, 11.3 roubles a pair from Hungary, 11.66 roubles a pair from Bulgaria and 12.08 roubles a pair from Czechoslovakia. The differences in unit values could, of course, reflect differences in the quality of imports from the respective market as described in Chapter 3. However, it is noticeable that the average price of EU imports of footwear from CMEA countries is one of the few items that was comparable with both extra-EU and intra-EU prices. The net effect is that in 1988, one tonne of Soviet crude oil exchanged for 4–5 pairs of footwear in trade with Austria and Yugoslavia compared with 10–13 pairs in trade with CMEA partners. Consequently although it appears that the official Soviet/CMEA exchange rate of TR:$1.646 overvalues Soviet exports of crude oil to CMEA partners in relation to world markets in 1998, it does not appear to provide such a substantial overvaluation of trade in footwear. This implies that the exchange rates used in Hungarian, Czech and Polish data may actually undervalue exports of consumer goods to the Soviet Union and the amount of resources that have gone into their production.

Nevertheless, there is general agreement that the official exchange rate between the transferable rouble and the dollar overvalues intra-CMEA trade in relation to extra-CMEA trade and results in an artificially high ratio of intra-CMEA trade in comparison with trade between the CMEA and the rest of the world in official CMEA statistics (Tarr, 1992, 25–6). The Economic Commission for Europe (ECE) has attempted to overcome the problem of multiple cross-rates by re-evaluating the foreign trade data of the CMEA economies on the basis of a uniform rouble-dollar exchange rate derived from the Hungarian cross-rate between the transferable rouble and the dollar (TR1.00:$0.52) which was felt to reflect market forces more accurately. Values of trade between CMEA partners in Tables 4.1–4.8 expressed in ECUs have been derived from the ECE cross rate for the dollar to the transferable rouble. These have then been converted into ECU at the 1988 exchange rate of ECU1.00:$1.182 to facilitate comparison with other Tables in the book. This method of evaluation reduces the value of intra-CMEA trade by 3.17 fold compared with data based on the official CMEA exchange rate for the transferable rouble. Estimates of the value of east European trade with the USSR in Tables 4.1–4.8 have been initially derived from Soviet trade data to ensure consistency in the reporting of trade with individual countries. These have been converted into ECU using the formula outlined above. The arguments above however, suggest that this may underestimate the value of certain east European exports of manufactured goods to the Soviet Union in general and for consumer goods in particular.

Tables 4.1 and 4.2 indicate that the ECE re-evaluation reduces the quantitative importance of intra-CMEA exports and imports as a proportion of total imports and exports for each east European economy by approximately 25 percentage points compared with the official CMEA estimates. This implies that intra-CMEA trade accounted for less than half of the total exports of the east European economies in 1988 compared with official estimates which placed them as closer to two-thirds

Table 4.1   East European exports by area of destination in 1988

	Destination (ECU billion)			CMEA as per cent of total exports			
	Total	of which		Official	Adjusted		
	World	CMEA*	USSR		Total	USSR	eastern Europe
Bulgaria	6.38	3.89	3.02	81.6	60.9	47.3	13.6
Czechoslovakia	10.47	5.29	2.99	74.2	50.5	28.6	21.9
GDR	13.38	5.52	3.09	66.0	41.3	23.1	18.2
Hungary	8.46	3.77	2.18	70.4	44.6	25.7	18.9
Poland	12.32	5.30	3.13	68.7	43.0	25.4	17.6
Romania	7.59	1.93	1.07	na	25.4	14.1	11.3
Total	58.62	25.73	15.48		43.9	26.4	17.5

*Excluding Mongolia, Cuba and Vietnam.
Sources and notes: for Tables 4.1 and 4.2: columns 1 and 2 from Economic Commission for Europe. Economic Bulletin for Europe, vol. 44, 132–3, converted into ECU at ECU:$1.182. Data for trade with the USSR have been estimated from Soviet data in Vneshniye Ekonomicheskiye Svyazi za CCCR, 1988, converted into dollars at the imputed ECE cross-rate of Transferable Rouble = ECU 0.44. The official CMEA estimate is provided in the CMEA Statistical Yearbook (Statisticheskii Yezhegodnik Stran-Chlenov SEV, 1989) and is based on the official TR:$ rate.

Table 4.2   East European imports by area of origin in 1988 ($ billion)

	Destination (ECU bn)			CMEA as per cent of total imports			
	Total	of which		Official	Adjusted		
	World	CMEA*	USSR		Total	USSR	eastern Europe
Bulgaria	6.88	3.44	2.68	74.3	50.1	39.0	11.1
Czechoslovakia	10.30	5.11	2.81	73.1	49.6	27.3	22.3
GDR	14.06	5.36	3.16	64.7	38.1	22.5	15.6
Hungary	7.93	3.47	1.97	69.5	43.7	24.9	18.8
Poland	10.99	4.73	2.77	68.6	43.0	25.1	17.9
Romania	4.53	1.89	1.03	na	41.6	22.7	18.9
Total	54.69	24.00	14.42		43.9	26.4	17.5

Sources and notes: see Table 4.1.

of total exports. The Soviet Union accounted for 60 per cent of east European exports to and imports from the CMEA, while Bulgaria and Czechoslovakia were the economies that were most dependent on Soviet and CMEA markets and imports.

4.6   The structure of intra-CMEA trade at the end of the communist era

Table 4.3 shows the commodity composition of Soviet exports to and imports from eastern Europe in 1988. 1988 has been chosen as it represents the last year before intra-CMEA trade flows were disrupted by the collapse of communism in eastern Europe. Intra-CMEA trade flows in 1988 were similar to those of 1986 and 1987. The commodity structure has been measured in intra-CMEA prices and consequently includes any distortions in relative prices contained in those figures. The interpretation of the data is further complicated by the fact that no commodity specification is provided for 16.5 per cent of the value of Soviet imports from eastern Europe and for 17.2 per cent of Soviet exports to eastern Europe. A cross-check with east European statistics on trade with the Soviet Union indicates that the majority of unspecified imports and exports fell into the category of machinery and equipment. This trade

Table 4.3   Soviet trade with eastern Europe, 1988 (ECU million)

	Soviet Imports		Soviet Exports		East European surplus/deficit	
	ECU mn	%	ECU mn	%	ECU mn	%
Total	15 487	100.0	14 431	100.0	+1 056	–
Machinery & equipment	8 480	54.7	2 412	16.7	+6 068	+38.0
Fuels & energy	292	1.9	7 011	48.6	−6 719	−46.7
Ores & metal products	378	2.4	1 059	7.3	−681	−4.9
Non-metallic minerals	40	0.3	92	0.6	−52	−0.3
Chemicals	373	2.4	355	2.5	−18	−0.1
Wood & paper products	40	0.3	347	2.4	−307	−2.1
Cotton & fibres	11	0.1	242	1.7	−231	−1.6
Materials for foodstuffs	169	1.1	7	–	+162	+1.1
Food products	710	4.6	70	0.5	+640	+4.1
Consumer goods	2 359	15.2	307	2.1	+2 052	+14.1
Other specified	80	0.5	29	0.2	+51	+0.4
Unspecified	2 555	16.5	2 500	17.3	+55	−0.7

Notes: Data have been estimated from Soviet export data to individual CMEA countries. Figures include the German Democratic Republic. Prices are actual intra-CMEA prices for 1988. These have been converted from valuta roubles into $ECU using the formula described in the text.
Source: Vneshniye Ekonomicheskiye Svyazi SSSR v 1988. Moscow, 1989.

normally included trade in strategic equipment including armaments and nuclear equipment which were normally conducted by bodies separate from the ministry of foreign trade. Unspecified trade may also include deliveries related to specific CMEA joint construction projects and trade in non-ferrous metals. An additional problem is that Soviet and east European trade data provide few details about the composition of trade flows of sub-categories *within* product categories (for example, rolled ferrous metals). This prevents the detailed estimation of intra-industry trade between CMEA partners and of the factor content and technological level of intra-CMEA trade that is comparable with western trade data.

Nevertheless, some clear principles of intra-CMEA trade can be observed from the aggregated data. The Soviet Union was a major net exporter of fuel and energy to eastern Europe which accounted for 48.6 per cent of the value of East European imports from the Soviet Union in 1988. This figure has been affected by the relationship between prices for energy relative to prices for manufactured goods in intra-CMEA trade. The combination of falling oil and gas prices and a fall in the volume of Soviet oil and gas exports to eastern Europe in the late 1980s meant that the share of Soviet energy exports was lower in 1988 than in the second half of the 1970s and the first half of the 1980s, when it regularly exceeded 50 per cent. Imported energy from the Soviet Union accounted for 31 per cent of all east European energy consumption (measured in thousand barrels a day oil equivalent) in 1987, ranging from 68 per cent of total consumption in Bulgaria to 13 per cent in Poland, net of Polish energy exports to the USSR (Smith, 1995). The east European economies as a whole were also net importers of iron ores and metallurgical products, wood and paper products and cotton and cotton fibres. The east European economies were substantial net exporters of machinery and equipment (which included transport equipment but contained no record of east European exports of cars) which accounted for 54.7 per cent of Soviet imports from eastern Europe. Industrial consumer goods (principally textiles and clothing, furniture, footwear, medicines, soaps and cosmetic goods and a relatively small proportion of electrical and white goods) accounted for 15.2 per cent of Soviet imports and materials for the production of foodstuffs (including grain, tobacco and seeds) and food products (including meat, fruit and vegetables, cigarettes and alcoholic and non-alcoholic drinks) together accounted for a further 5.7 per cent of east European exports to the Soviet Union.

It is apparent from Table 4.3, that intra-industry trade was relatively unimportant in east European trade relations with the Soviet Union, even at this high level of aggregation of both products and trade partners. The last column in Table 4.3 subtracts the proportion of east European imports from the Soviet Union in each of these categories from the proportion of east European exports to the Soviet Union in the same category to provide a crude commodity balances for east European trade with the Soviet Union. As price distortions in intra-CMEA trade were more pronounced *between* major categories of goods, than *within* categories, the *sign* of the balances, as opposed to their size, will be relatively unaffected by the use of intra-CMEA prices. Table 4.3 indicates particularly high levels of specialisation between the Soviet Union and eastern Europe as a whole with differences between the proportion of exports and imports of 38 per cent for machinery and equipment and 14.1 per cent for light consumer goods and minus 46.7 per cent for fuel and energy. The problem of the exchange rates used to value imports and exports is not important in this case as they will both be affected pro rata.

Tables 4.4 and 4.5 provide a breakdown of the structure of Soviet imports from, and exports to, individual CMEA economies. The unusual feature of Table 4.4 is the relative lack of differentiation between the structure of Soviet imports of manufactured goods from the individual countries. The chief exceptions were that Poland was the only (recorded) East European exporter of fuel and energy to the Soviet Union accounting for all Soviet imports of ECU 292 million in this sector, while Bulgaria and Hungary were the only significant exporters of foodstuffs. Machinery and equipment predominated in Soviet imports from each country, ranging from 47.2 per cent from Poland to 64.4 per cent from

Table 4.4   Soviet imports from eastern Europe in 1988: commodity structure (ECU mn)

	Bulgaria	Czechoslovakia	GDR	Hungary	Poland	Romania
Machinery	1 665	1 731	1 991	1 094	1 477	522
Food and materials	418	18	0	334	48	59
Industrial consumer goods	337	490	496	297	509	230
Unspecified	546	475	397	372	436	130
Total (including others)	3 024	2 999	3 091	2 175	3 128	1 069

Sources and notes: see Table 4.3.

Table 4.5 Soviet exports to eastern Europe in 1988: commodity structure (ECU mn)

	Bulgaria	Czechoslovakia	GDR	Hungary	Poland	Romania
Machinery	665	371	518	290	402	166
Energy	1 136	1 561	1 519	923	1 363	509
Iron and steel	101	158	374	105	202	119
Chemicals	33	76	43	106	70	27
Minerals, wood, paper and cotton	75	74	180	158	128	67
Unspecified	615	490	502	323	436	135
Total (including others)	2 681	2 809	3 165	1 973	2 771	1 031

Sources and notes: see Table 4.3.

Czechoslovakia.[2] Each individual country was also a significant exporter of industrial consumer goods, ranging from 11.1 per cent of the total for Bulgaria to 21.5 per cent for Romania. The structure of exports of machinery and equipment and industrial consumer goods is examined in Tables 4.7 and 4.8. Energy predominates in Soviet exports to each east European country, with smaller shares of metallurgical goods and industrial inputs in the form of minerals, wood, paper and cotton.

Table 4.6 provides estimates of broad commodity balances for the trade of each individual CMEA economy with the Soviet Union.[3] These figures demonstrate the 'radial' pattern of Soviet economic relations with eastern Europe, as each separate CMEA economy was a substantial net exporter of machinery and equipment and industrial consumer goods and a substantial net importer of fuel and energy. There is some degree of variation in the level of net exports of machinery and equipment with higher levels for the more industrialised economies of the GDR and Czechoslovakia. Nevertheless, the recorded rates of over 30 per cent for the net exports of machinery and equipment by the more agrarian economies of Romania and Bulgaria are high.

4.7 Soviet–east European trade in specific goods

4.7.1 East European exports of machinery and equipment to the Soviet Union in 1988

Machinery and equipment (not including those items of machinery and equipment that cannot be identified in Soviet statistics) accounted for

Table 4.6   Commodity balances in individual east European countries' trade with the Soviet Union in 1988

	Bulgaria	Czechoslovakia	GDR	Hungary	Poland	Romania
Machinery & equipment	+30.2	+44.5	+47.3	+35.6	+32.7	+32.7
Fuels & energy	−42.4	−55.6	−48.0	−46.8	−39.9	−49.3
Iron ore & steel products	−3.6	+0.7	−11.5	−4.5	−4.5	−5.3
Non-metallic minerals	−0.2	−0.4	−0.7	−0.7	+0.4	−1.9
Chemicals	−0.3	−0.2	+3.3	−3.1	−1.1	−
Wood & paper products	−1.4	−0.8	−2.8	−5.3	−1.7	−0.8
Cotton & fibres	−1.1	−1.4	−1.6	−1.9	−2.2	−1.7
Materials for foodstuffs	+1.9	−0.2	−0.1	+3.8	+0.5	+0.1
Food products	+11.9	+0.4	−0.1	+10.9	−	+4.7
Light consumer goods	+9.1	+14.3	+15.3.	+11.3	+11.5	+21.5
Unspecified	−4.9	−1.6	−3.2	+0.7	−4.5	−0.8

Notes: Commodity balances have been estimated by subtracting the share of imports in each product category as a percentage of total imports from share of exports in the same product category as a percentage of total exports.
Sources and methods: as for Table 4.3.

54.7 per cent of east European exports to the Soviet Union in 1988 and amounted to ECU 8480 million. Just under a quarter of this figure (22.41 per cent) cannot be identified more precisely from Soviet data. 25.8 per cent of east European machinery exports to the Soviet Union in 1988 can be clearly identified as inputs to investment in heavy industry; metal working equipment (5.9 per cent), power and electrical equipment (5.99 per cent), equipment for the mining and energy industries, including equipment for the exploration, extraction, refining and transportation of oil (6.99 per cent) and hoisting and conveying equipment including cranes, winches, and fork-lift trucks and similar industrial equipment (6.91 per cent). Transport equipment, predominantly for transporting freight and public passenger transport, with a relatively low proportion for private transport, accounted for a further 22.67 per cent with motor transport (principally lorries and buses) accounting for 9.66 per cent railroad equipment for 6.22 per cent and boats and maritime equipment for 5.15 per cent Equipment for specific industries took up a further

15 per cent. More sophisticated items including medical instruments and laboratory and computing equipment accounted for only 11.47 per cent of east European exports of machinery and equipment to the Soviet Union in 1988.

Although all countries, except Romania, exported items in at least 14 of the 15 categories identified in Table 4.7 (with even landlocked countries like Hungary and Czechoslovakia exporting maritime equipment) some significant levels of export specialisation can be observed for individual countries. Bulgaria was specialised in hoisting and conveying equipment (particularly concentrated in the production of fork-lift trucks) and computing equipment. These two items together were worth ECU 829 million, at the revised exchange rate described above, and accounted for 49.78 per cent of Bulgarian exports of machinery and equipment to the USSR and were equivalent to 13 per cent of total Bulgarian exports to all markets in 1988. Czechoslovakia had an export specialisation in motor transport equipment (especially lorries), railroad equipment and equipment for the sewing and textile industries. The GDR was specialised in the production of boats and maritime equipment and metalworking equipment. Somewhat surprisingly, in view of the GDR's high level of industrialisation, and the fact that it was not a major food producer, the GDR also had a high level of specialisation in the export of agricultural equipment and equipment for the food industry. Hungary was specialised in the production of medical instruments and laboratory equipment and motor transport equipment. The latter, which accounted for 36.9 per cent of Hungarian exports of machinery and equipment to the Soviet Union, was made up of coaches and spare parts worth ECU 390 million. Poland was specialised in the production of power and electrical equipment, boats and maritime equipment and medical instruments and laboratory equipment. Finally, Romania, which did not fully participate in a number of CMEA cooperation ventures and had the lowest absolute and per capita level of trade with the Soviet Union of all the east European economies, had the highest degree of specialisation in trade in machinery and equipment with the Soviet Union. Romanian exports to the Soviet Union were concentrated in 9 of the 15 product categories itemised in Table 2.5 with exports of equipment for the mining and energy industries composed of equipment for drilling and geological prospecting worth ECU 187 million accounting for 37.76 per cent of Romanian exports of machinery and equipment and railroad equipment accounting for a further 8.20 per cent.

Table 4.7 Soviet imports of machinery and equipment from eastern Europe in 1988

	Bulgaria	Czechoslovakia	GDR	Hungary	Poland	Romania	Total
Value (ECU mn)	1 665	1 731	1 991	1 094	1 477	522	8 480
Imports of M&E as % of imports	55.1	57.7	64.4	50.3	47.2	48.8	54.7
Specific items as a percentage of imports of machinery and equipment:							
Metalworking machine tools	7.51	5.05	9.72	1.51	3.76	5.22	5.95
Power & electrical equip	4.01	6.49	3.95	1.92	13.67	5.25	5.99
Equip for mining and energy[a]	1.53	5.82	6.26	1.14	8.97	37.76	6.99
Hoisting and conveying equip[b]	19.71	3.38	4.90	2.65	4.95	0	6.91
Equip for food industry	1.65	3.75	6.56	4.37	1.25	0	3.35
Equip for textile and sewing	1.48	7.88	2.07	1.59	1.93	0.93	2.98
Equip for chemical industry	0.31	3.54	3.56	0.82	2.30	3.28	2.33
Road building equip	0.30	2.11	2.12	0.09	3.66	0	1.64
Medical instruments and laboratory equip	1.49	2.47	4.78	7.70	6.17	0	3.99
Computing equip	30.07	0	2.75	0	5.35	0	7.48
Agricultural equip	4.70	4.62	10.98	4.67	2.88	6.54	5.95
Railroad equip	0	10.03	7.98	1.28	5.83	18.15	6.22
Motor transport equip[c]	1.40	17.92	0.22	36.95	5.25	0	9.66
Boats & maritime equip	2.29	1.81	10.02	1.21	8.20	6.47	5.15
Other specified[d]	1.00	3.28	4.40	1.94	2.93	4.25	3.00
Total specified	77.45	78.15	80.25	67.84	77.10	87.85	77.59
Unspecified	22.55	21.85	19.75	32.16	22.90	12.15	22.41

Notes: a) including pumps and compressors and pipeline equipment; b) predominantly fork-lift trucks and similar equipment; c) Predominantly lorries and buses, but also including a small amount of cars and motorbikes; d) including equipment for firefighting and the printing industry. Machinery and equipment has been measured according to the CMEA definition which includes items of transport equipment that are not included under that heading in conventional international statistical systems.
Source: Estimated from Vneshniye Ekonomicheskiye Svyazi SSSR v 1988, Moscow, 1989; and Vneshniye Ekonomicheskiye Svyazi SSSR v 1989, Moscow, 1990.

4.7.2 East European exports of industrial consumer goods to the Soviet Union in 1988

Soviet imports of industrial consumer goods (according to the CMEA nomenclature which includes some items that would be classified as chemicals in the SITC nomenclature) from eastern Europe amounted to ECU 2359 million in 1998 and accounted for 15.2 per cent of Soviet imports. Table 4.8 shows that Soviet imports of light consumer goods (clothing, textiles, footwear, household ware and furniture) amounted to ECU 1451 million and were a significant component of Soviet imports from each east European country. Soviet imports of household appliances and electrical goods from eastern Europe were either negligible or non-existent.[4] Despite Czechoslovakia's relatively high level of industrialisation in the inter-war period, household appliances and electrical goods accounted for only 1.4 per cent of Soviet imports of industrial consumer goods from Czechoslovakia, while Soviet imports of clothing, textiles, footwear and furniture accounted for 89 per cent of imported consumer goods from Czechoslovakia. Similarly, household appliances accounted for only 4 per cent of Soviet imports of consumer goods from the GDR. The Soviet Union provided an important market for Bulgarian, Hungarian and Polish exports of chemical-based consumer goods. Soviet imports of medicines and cosmetics amounted to ECU 199 million and accounted for 59.1 per cent of imported consumer goods from Bulgaria. Soviet imports of medicines from Poland amounted to ECU 244 million and accounted for 48 per cent of imports of consumer goods and to ECU 121 million from Hungary to account for 41 per cent of imported consumer goods. Romania had the least sophisticated structure of exports of consumer goods to the Soviet Union with clothing, furniture and footwear accounting for 94 per cent of the total under this heading.

4.8 Trade with the industrialised west and the EU

4.8.1 The strategy of import-led growth

By the late 1960s, central planners and politicians in eastern Europe realised that their isolation from western markets and multinational investment and technology was contributing to a growing gap between the technological level and quality of manufactured products in eastern and western Europe. Consequently, reformist economists advocated dismantling the foreign-trade monopoly, exposing domestic enterprises to foreign competition and world market prices and the

Table 4.8 Soviet imports of consumer goods from eastern Europe in 1988

	Bulgaria	Czechoslovakia	GDR	Hungary	Poland	Romania	Total
Total(ECU mn)	337	490	496	297	509	230	2 359
Light consumer goods	136	436	313	163	184	219	1 451
of which:							
Clothing	84	210	222	110	120	137	883
Footwear	24	143	8	49	64	27	315
Household ware	0	17	14	0	0	0	31
Furniture	28	66	69	4	0	55	222
Chemical-based	199	32	112	125	278	2	748
of which:							
Medicines	100	32	49	121	244	2	548
Cosmetics, soaps, perfumes	99	0	63	4	34	0	200
Other	2	15	44	8	31	8	108
of which:							
Electrical goods & appliances	0	7	27	1	16	1	52

Notes and sources: estimated from *Vneshniye Ekonomicheskiye Svyazi za 1988*. Conversion into ECU as outlined in the text.

import of western machinery and equipment.[5] As the isolation from western technology was reinforced by western controls over exports of technology with potential military applications to communist countries, this would necessarily involve attempts to improve political relations between communist and non-communist states. Most reformers envisaged that a radical reform of the domestic system of planning would be complementary to a reform of the system of foreign trade. In practice, the backtracking on reforms after the Warsaw Pact invasion of Czechoslovakia in 1968 resulted in the implementation of limited domestic economic reforms and strictly controlled attempts to develop contacts with western multinationals in an attempt to acquire western technology as part of a strategy which became known as 'import-led growth.' (Smith, 1983; Zloch-Christy, 1988).

Contacts with western multinationals ranged from cooperation and subcontracting arrangements, the purchase of licences to the development of joint ventures with western firms on east European territories. Western companies were normally restricted to minority shareholdings in joint ventures which gave them limited ownership rights and restricted control over the detailed operation of the venture and consequently made them less attractive to western companies than investing in wholly-owned subsidiaries in newly-industrialising economies. As a result, the east European economies resorted to outright purchases of machinery and equipment and complete installations in order to modernise and extend the industrial capital stock. This forced them to take credits in convertible currencies. It was expected (or hoped) that credits would be repaid from increased hard-currency revenues which would be generated by exports produced by the imported equipment. However, hard currency exports from eastern Europe grew more slowly than imports throughout the 1970s, resulting in widening current account deficits and growing problems of indebtedness. These were greatest in Poland and Romania and had become critical by the late 1970s. At this stage east European imports of machinery and equipment from OECD economies had reached annual levels of approximately $6 billion. In 1979 the CMEA-5 together incurred annual trade deficits of $3.5 billion (equivalent to ECU 2.55 billion at the then exchange rate) and current account deficit of $5.8 billion (ECU 4.23 billion) in convertible currency. The gross indebtedness of the CMEA-5 had reached $55.2 billion ($51.9 billion net of foreign exchange reserves) by the end of 1981 (UNECE, 1992, 138–9).

4.8.2   Problems of indebtedness

Poland and Romania were forced to reschedule their external debts in 1981, while Hungary narrowly avoided rescheduling. This, combined with the deterioration in the international financial climate in 1982 and the desire by financial institutions to restore liquidity following the Latin American debt crisis, resulted in to a loss of confidence in lending to eastern Europe and the withdrawal of new credit facilities by western financial institutions. The failure of the east European economies to increase exports to OECD markets meant that they could only respond to the reduced availability of external credits by restricting imports from the OECD. This was partially achieved by reducing imports of machinery and cutting back investment as each of the East European economies managed to run a balance of trade surplus in the period 1982–4. This policy was modified in 1985 after Gorbachev came to power in the Soviet Union. The CMEA-5, with the exception of Romania – which continued to pursue a policy of rapid debt-repayment – relaxed controls over imports, but continued to experience difficulties in expanding convertible currency imports. By the time of the collapse of communism in 1989 gross hard currency debt of the CMEA-5 had reached $79 billion (UNECE, 1992).

The question of why the East European economies failed to generate a sufficient volume of exports to service their debt is of major significance for assessing the ability of the CEE economies to withstand competitive pressures from EU producers. For seventeen consecutive years from 1965 to 1981, the centrally planned economies of the Soviet Union and eastern Europe (including Yugoslavia and the GDR) incurred deficits in their trade balances and current accounts conducted in convertible currencies which resulted in a combined gross debt of $119 billion by the end of 1981. The reduced availability of external credits required them to eliminate trade deficits in the early 1980s which was largely achieved by restricting imports. Nevertheless, the combined gross convertible currency debt of the communist economies had reached $189 billion by the end of 1989.

4.8.3   The causes of difficulties in exporting to the West

Were the current account deficits incurred by east European economies in the communist era caused by lax macroeconomic policies which, combined with the availability of external finance, sucked in excessive imports and allowed potential exports to be diverted to domestic markets, or were they caused by more fundamental, structural problems? The proposition that balance of payments problems can arise as a result of structural problems which contribute to an 'inability to export' is based on the analysis developed by Hirschmann (1975) to explain the balance of payments problems encountered by developing economies, particularly in Latin America in the 1970s, and which was adapted to analyse the indebtedness of east European economies under communism in the 1970s by Wiles and Smith (1978). The essence of this argument is that country-specific, or region-specific, institutional factors may combine to prevent an economy from responding quickly to innovations that are taking place in technologically more advanced economies of the world. This will prevent those economies from adapting to changing conditions on world markets and from producing goods that meet the quality standards and product specifications that are required to meet the demands of consumers in high-income economies. The technology, or quality, gap between a country's domestic production standards and those of world markets may be so large that foreign demand for the country's products becomes price-inelastic with the result that the country is unable to increase export revenues by reducing the price of exports. Under these circumstances, devaluation becomes ineffective as a means for curing external deficits. Although this argument is feasible, it is likely to apply to special cases only, and may arise from circumstances where governments pursue protectionist policies, to import-substituting policies, in response to an initial balance of payments problem (see Holzman, 1979).

The burden of explanation for the poor export performance in western markets by east European economies under communism, rested on the nature of the centrally-planned system and the state monopoly of foreign trade in particular, which isolated domestic producers from foreign competition and provided them with a guaranteed domestic and CMEA markets for obsolete products. These problems were compounded in the CMEA by the adoption of Stalinist patterns of production which emphasised quantity over quality, and gave priority to the development of civil and mechanical engineering industry over electronic engineering, and by socialist patterns of consumption which attached low priority to variety and sophistication in the production of consumer goods. The knowledge that enterprises faced soft budget constraints (the state would normally cover the losses of losing enterprises by means of subsidies, non-repayable credits and adjustments to tax demands) and would not face bankruptcy in the event of failure to meet the demands of consumers provided them with little or no incentive to adapt to the more stringent demands of foreign markets, or to innovate to produce new and improved products, or to introduce cost-saving processes.

4.9   The structure of EU imports from central and south-east Europe at the end of the communist era

4.9.1   The commodity structure of EU imports from central and south-east Europe

Table 4.9 shows the structure of EU-12 imports from the CMEA-5 in 1988 together with data on extra-EU trade for purposes of comparison. Extra-EU data refer to the trade of the current EU-15 members. These data are not perfectly comparable with data for the EU-12, but have the advantage that they do not include trade between the EU-12 and the higher income economies of Austria, Sweden and Finland in extra-EU trade. The export specialisation index for each country's trade in any given sector can be inferred directly from Table 4.9 by comparing the figure in any given cell with the corresponding row in the final column. The breakdown between primary goods and manufactured goods (36:64) in the exports of the CMEA-5 is not significantly different from that in extra-EU trade (33:67). Agricultural products took up a larger proportion of EU imports from the CMEA-5 (19.8 per cent) than from the world as a whole (14.6 per cent) while CMEA-5 exports of fuel and energy were only important for Poland and Romania. Romanian exports of energy largely consisted of refined oil products which were derived from imported crude oil. Machinery and equipment accounted for a significantly smaller share of EU imports from the CMEA-5 (11.8 per cent) than in EU imports from the world as a whole (29.6 per cent), with Romania having the lowest share. The low share of CMEA machinery exports was offset by larger shares of light consumer goods (principally clothing, furniture and footwear) and minerals and metals (pottery, glassware, iron and steel manufactures and non-ferrous metals) in the exports of the CMEA-5. The share of light consumer goods in the exports of each of the CMEA-5 was substantially higher than the share of these goods in EU imports from the world, while the commodity composition of light consumer goods indicates that each of the CMEA-5 had an export specialisation in clothing and furniture and (with the exception of Bulgaria) in footwear in trade with the EU. This reflects the high proportion of these goods in exports to the Soviet Union. Czechoslovakia had a relatively low proportion of exports of light consumer goods to the EU, compared to their weight in exports to the Soviet Union.

Table 4.9 The structure of EU-12 imports from the CMEA-5 in 1988

	Bulgaria	Czechoslovakia	Hungary	Poland	Romania	Total	Extra-EU-15
Primary goods	**39.1**	**27.1**	**36.8**	**41.0**	**35.6**	**35.9**	**33.0**
Agriculture	26.3	14.0	30.8	24.7	6.2	19.8	14.6
Fuels, Energy, Minerals	10.7	9.5	5.1	16.0	28.5	14.8	16.1
Manufactures	**60.9**	**72.9**	**63.2**	**59.0**	**64.4**	**64.1**	**67.0**
Machinery	10.4	14.3	12.1	13.8	6.6	11.8	29.6
Chemicals	**16.8**	**12.0**	**10.3**	**5.8**	**5.1**	**8.4**	**6.9**
organic	9.3	6.0	4.6	1.5	2.7	3.7	1.9
medicines	1.3	0.5	0.5	0.1	0.2	0.4	1.0
fertilisers	1.7	0.3	1.3	0.6	1.4	0.9	0.3
plastics	1.6	3.2	2.7	1.0	0.3	1.7	1.3
Light consumer goods	**10.3**	**9.3**	**18.2**	**14.3**	**30.7**	**17.4**	**6.1**
Clothing	7.4	4.9	12.7	9.2	15.8	10.4	4.4
Footwear	0.3	1.4	2.7	1.7	2.1	1.9	0.8
Furniture	1.9	2.2	2.3	3.0	12.6	4.7	0.6
Minerals & metals	**12.5**	**19.6**	**12.2**	**18.3**	**15.2**	**16.4**	**8.8**
Mineral goods	1.5	5.5	2.0	2.1	2.7	2.9	1.8
Iron & steel	7.6	13.8	7.5	6.8	6.6	8.4	3.7
Non-ferrous Metals	3.4	0.3	2.7	9.5	5.9	5.1	3.3
Total (ECU mn)	**455**	**2 191**	**2 148**	**3 292**	**2 227**	**10 312**	**360 879**

Source: estimated from COMEXT database.

If Czechoslovakia is ignored, the country rankings according to the share of total exports of light consumer goods is the same in both the Soviet and the EU markets, with Romania having the highest proportion of exports of light consumer goods to both markets, followed by Hungary and Poland with Bulgaria in last place. Although chemicals occupied a greater share of EU imports from the CMEA-5 than from the outside world, this was largely concentrated on low-value organic chemicals and fertilisers while the CMEA-5 had little success in penetrating EU markets for higher-value products such as pharmaceuticals and medicines, although these products were a significant component of exports to the Soviet Union for Poland, Hungary and Bulgaria in particular.

The share of minerals and metals in CMEA-5 exports to the EU was nearly double that of the share of minerals and metals in EU imports from the rest of the world, with iron and steel goods over twice the EU average. Each CMEA country showed an export specialisation in iron and steel products, with Czechoslovakia the most prominent.

4.9.2   The factor content and technological level of CMEA exports to the EU in the communist era

The factor content and technological level of CMEA-5 exports to the EU is shown in Table 4.10 which uses the Wolfmayr-Schnitzer classification described in Chapter 3. This confirms that the CMEA-5 share of exports of human capital-intensive goods was less than half the share of these items in total EU-12 imports from outside the EU, while exports of human capital-intensive goods embodying high technology were less than one-sixth of the share of these products in EU imports. The relatively low level of high-technology products was largely the result of the failure of the CMEA-5 economies to produce high quality data-processing equipment and other electronic items and items of precision engineering. However, it also reflected the failure to penetrate EU markets for chemical products embodying high-technology, including pharmaceuticals and insecticides. The relatively low share of medium technology goods in the human capital-intensive sector (principally machinery and other chemical products) was also disappointing in view of the high share of these goods in exports to the Soviet Union. The low share of human capital-intensive exports was offset by substantially higher export shares of labour-intensive products (including clothing, footwear and furniture) and resource-intensive products (chiefly minerals, metals and fertilisers).

Differences between the export structures of the individual CMEA-5 economies are reflected in the technological level and factor-content of

Table 4.10 The factor content of CMEA exports of manufactured goods to the EU-12 in 1988 (percentage of exports of manufactured goods)

	Bulgaria	Czechoslovakia	Hungary	Poland	Romania	Total	Extra EU-15
Human Capital-Intensive	32.56	27.35	29.34	31.86	12.94	26.12	55.97
High technology	6.51	2.40	4.95	2.27	1.42	2.83	18.94
labour-intensive	2.43	1.15	3.75	1.56	0.67	1.75	11.30
capital-intensive	4.08	1.25	1.20	0.70	0.75	1.08	7.64
Medium technology	21.42	21.05	21.39	19.35	9.71	18.14	30.54
labour-intensive	13.18	8.62	11.71	7.35	4.76	8.21	19.71
capital-intensive	8.24	12.43	9.68	11.99	4.95	9.92	10.83
Other	4.62	3.91	3.00	10.25	1.80	5.16	6.50
Labour-Intensive	27.78	33.68	44.31	34.55	56.18	40.82	25.57
Resource-Intensive	39.67	38.97	26.35	33.59	30.88	33.05	18.45
Total (mn ECU)	262	1 562	1 328	1 913	1 430	6 495	239 010

Source: estimated from COMEXT database.

CMEA exports of manufactured goods to the EU which are shown in Table 4.10. Although Bulgaria was regarded as one of the least developed of the CMEA economies, it actually had the highest proportions of both human capital-intensive and high-technology exports of the CMEA-5 economies. However, this was recorded on a very low volume of exports of manufactured goods. Bulgarian exports of human capital-intensive goods embodying high technology only amounted to ECU 17 million and were largely concentrated in the chemicals sector, particularly pharmaceuticals which was also a major source of exports to the Soviet Union. Despite the high proportion of Bulgarian exports of data processing equipment to the Soviet Union, exports of these items to the EU only amounted to ECU 0.5 million. Romania is the clearest outlier in Table 4.10 with very low shares of human capital-intensive products and goods embodying high technology and correspondingly high shares of labour-intensive goods.

4.10   Relative unit values of EU imports from central and south-east Europe in the communist era

A final indication of the quality of CMEA-5 exports to the EU in the communist era can be obtained by comparing the unit values of CMEA exports to the EU with those of other exporters to the EU. Table 4.11 compares the ratio of the unit value (measured in ECUs per tonne of exports) of EU imports from the CMEA exports to EU imports from the rest of the world of selected manufactured goods. The unit value of EU imports of manufactured goods from the CMEA-5 at ECU 746 per tonne was only 29.1 per cent of the unit value of total EU manufactured goods from outside the EU of ECU 2565 per tonne, resulting in a unit value index (UVI) of 0.291. It was argued in Chapter three that a low unit value for a product grouping can be attributed to the a lower share of high-value exports within that category and/or to lower prices obtained for similar products. The latter is taken to indicate that consumers consider the product to be of a lower quality than that of competing goods which is reflected in a lower price. The lower aggregate unit value for CMEA exports partly reflects the low share of exports of high unit value goods such as aircraft, data processing equipment, telecommunications and other electronic equipment and pharmaceuticals which was reflected in the low relative shares of exports of human capital-intensive goods embodying high and medium technology. However, it also reflected a disturbing quality differential across a wide range of manufactured products. The CMEA-5 as whole recorded a UVI of greater than

Table 4.11 Ratio of unit values of EU-12 imports of manufactured goods from CMEA-5 to extra-EU imports in 1988

	Bulgaria	Czechoslovakia	Hungary	Poland	Romania	Total
Total manufactures	**0.235**	**0.265**	**0.352**	**0.295**	**0.281**	**0.291**
Chemicals	**0.462**	**0.651**	**0.484**	**0.395**	**0.258**	**0.444**
organic	0.571	0.608	0.532	0.645	0.341	0.525
inorganic	0.550	0.703	0.424	0.228	0.631	0.347
pharmaceuticals	0.320	1.421	0.865	0.303	0.097	0.360
fertilisers	0.633	0.521	0.953	0.874	0.774	0.800
plastics	0.553	0.695	0.792	1.203	0.771	0.681
Machinery	**0.202**	**0.196**	**0.267**	**0.185**	**0.203**	**0.204**
general purpose	0.212	0.236	0.263	0.186	0.266	0.229
computers, office equipment	0.138	0.220	0.137	0.054	0.624	0.109
telecoms, sound recording	0.769	0.184	0.236	0.160	0.150	0.187
road vehicles	0.667	0.396	0.421	0.403	0.422	0.405
Manufactured goods classified by material (SITC 6)						
Total	0.347	0.364	0.438	0.396	0.344	0.380
Mineral products	0.156	0.277	0.219	0.090	0.114	0.152
Iron and steel	0.540	0.627	0.606	0.600	0.609	0.610
Non-ferrous metals	0.666	0.465	0.792	0.938	0.714	0.833
Miscellaneous manufactured goods (SITC 8)						
Total	0.356	0.303	0.550	0.380	0.243	0.325
Clocks & watches	0.166	0.102	0.103	0.097	0.049	0.090
Clothing	0.760	0.846	1.532	1.207	0.926	1.090
Footwear	0.617	0.537	1.719	1.002	1.044	0.996
Furniture	0.606	0.619	0.801	0.549	0.603	0.608

Source: estimated from COMEXT database.

one in only 15 of the 166 three-digit SITC categories of manufactured goods which the EU imported from CMEA. Five of these were categories in which CMEA exports were below ECU 1 million and a further three were residual categories for goods that could not be classified elsewhere. The remaining items were either non-ferrous metals or goods associated with the clothing industry.

This is reflected in the selected indicators shown in Table 4.11. The relative standing of the countries for UVIs for total manufactures is affected by the commodity structure of exports. Czechoslovakia which had high shares of resource-intensive goods with a relatively low ratio of value to weight (organic chemicals, steel products) in its export structure recorded a low overall UVI which was below Romania which had a high share of clothing exports which have a medium to high value to weight ratio. This indicates that the aggregate figures need to be treated with some caution. However the picture that emerges from Table 4.11 is that the UVI gap is highest for human capital-intensive products (computers, telecommunications), then narrows for resource-intensive products, (iron and steel and non-ferrous metals) and closes for some labour-intensive exports (footwear), while the UVI of EU imports of clothing from CMEA is higher than the average for EU imports of clothing. There is also a major difference between the UVI gap for exports of miscellaneous manufactured goods as a whole (SITC 8) and for specific items such as clothing and footwear. This reflects the lower share of exports of high value goods in this category including photographic equipment, optical and surgical equipment and instruments and optical fibres, while items like clocks and watches were sold at the very bottom of the market range.

Conclusions

The tests analysed in this Chapter indicate that the imposition of the Stalinist model of industrialisation created major structural problems for the east European economies which inhibited their ability to export to the EU. These problems were reinforced by the protectionist, import-substituting system of foreign trade which required the east European economies to meet the demands of the CMEA market and the Soviet Union in particular. Although the east European economies were major net exporters of machinery and equipment to the Soviet Union this was concentrated on equipment for heavy industry including metallurgy, power generation, transportation and the extraction of minerals. The production and export of goods which embodied human capital-intensive processes including computer equipment, data processing and telecommunications was relatively neglected. The unit value of CMEA exports of specific items of machinery to the EU was, on average, more than 70 per cent below that of EU exports from the rest of the world, suggesting that they were of substantially lower quality than EU imports from other parts of the world. The east European economies were also major net exporters of consumer goods to the Soviet Union. However the structure of CMEA exports was geared towards a captive market in which the structure of demand was determined by central planners, not market forces. This resulted in an export structure that was geared towards relatively simple products including basic medicines, clothing, footwear and furniture. More sophisticated household appliances, and more expensive consumer goods including cars received low priority. This was reflected in the structure of exports of manufactured goods from the east European CMEA economies to the EU. At the end of the communist era, the export structure of the CMEA economies was biased away from human capital-intensive products which embodied high and medium technology towards resource-intensive products and labour-intensive products which were normally by-products of goods produced for the Soviet market. This indicates that the problems of convertible-currency indebtedness incurred by the east European economies in the 1970s and 1980s were not simply the result of loose macroeconomic policies which sucked in imports and diverted potential exports to domestic markets, but reflected deeper structural problems with microeconomic causes and that major industrial restructuring would be required before the CMEA economies could compete in the production of human capital-intensive goods embodying high and medium technology.

5
The Basic Features of Trade Relations between the EU and the Applicant States

5.1 Introduction

The collapse of communism in central and south-east Europe in 1989, followed by the dissolution of the Soviet Union in 1991 and the disbanding of the CMEA destroyed the rationale on which the trade relations between communist states had been constructed. This required the former CMEA economies and the Baltic states to seek new trade partners. Similarly, the secession of Slovenia from Yugoslavia in 1991–2, and the dislocation created by the outbreak of war in the former Yugoslav lands, required Slovenia to intensify its trade relations with non-Yugoslav states. The EU became an obvious target for the redirection of trade. The countries which now constitute the EU had been the major trade partners for each of the economies in the inter-war period and their geographic proximity and their relatively large per-capita incomes made them an attractive proposition for exporters seeking new markets. The growth of trade between the ten central and east European economies that are being considered for entry into the EU in the first two tiers of eastward enlargement (the CEE-10) and the current fifteen members of the EU (the EU-15) was substantial in the early 1990s. However, it has not been without problems. EU-15 exports to the CEE-10 have grown substantially faster than CEE-10 exports to the EU-15, resulting in the re-emergence of large trade deficits and problems of indebtedness. This raises an important question about the cause of deficits in CEE trade with the EU. Are they purely a transitional phenomenon related to the increased demand for capital for restructuring? Or, do they result from more fundamental problems arising from a structural inability to produce the more sophisticated goods that are demanded in EU markets?

This Chapter will present some basic data concerning the growth and commodity structure of trade between the EU-15 and the CEE-10 in the mid-1990s. It will examine the extent to which trade flows between the CEE-10 and the EU-15 contribute to a net inflow of capital goods to the CEE-10 which will increase their productive capital in the long term and how much they are being used to increase short-run domestic consumption. It will also examine the importance of outward processing trade to the CEE-10 economies. These primary data will be subjected to more detailed tests and analysis in Chapters 6 and 7.

5.2   The size and growth of trade between the EU-15 and the CEE-10

Tables 5.1–5.3 show the size and growth of the imports, exports and trade balances of the individual CEE-10 economies with the EU-15. Data on trade flows with the new independent states that emerged from the break-ups of the Soviet Union and Yugoslavia were subject to considerable statistical error in 1992 and should be treated with caution. The

Table 5.1   EU-15 exports to CEE-10, 1989–97 (ECU million)

	1989	1992	1993	1994	1995	1996	1997
Bulgaria	1 706	1 255	1 488	1 752	2 052	1 696	1 705
Czechoslovakia	2 902	7 483					
Czech			7 087	9 224	11 653	13 966	15 823
Slovakia			1 583	2 194	3 192	3 996	4 773
Hungary	3 793	5 284	6 447	8 066	8 728	9 989	13 417
Poland	4 594	9 219	11 114	12 317	15 294	19 827	25 186
Romania	744	1 988	2 513	2 895	3 794	4 437	4 996
Total CMEA-5	13 739	25 229	30 232	36 448	44 713	53 911	65 900
CMEA-5 as % extra-EU trade	3.52	6.13	6.41	6.96	7.81	8.64	9.07
Slovenia		1 382	3 618	4 348	5 178	5 376	6 291
Estonia		334	492	1 030	1 348	1 694	2 311
Latvia		251	415	719	940	1 107	1 516
Lithuania		365	545	856	1 016	1 451	2 121
CEE-10 total		27 561	35 302	43 401	53 195	63 539	78 139
CEE-10 as % extra-EU trade		6.70	7.49	8.29	9.30	10.19	10.75

Sources: 1989, 1992–6 from External and Intra-European Trade, Statistical Yearbook, 1958–1996, Luxembourg, 1997; 1997 estimated from Direction of Trade Statistics Yearbook, 1998. Data are taken from EU sources which differ from those in Tables 1.4 and 1.5, which are taken from CEE sources.

Table 5.2   EU-15 imports from CEE-10, 1989–97 (ECU million)

	1989	1992	1993	1994	1995	1996	1997
Bulgaria	587	977	1 014	1 419	1 836	1 701	2 107
Czechoslovakia	3 231	6 499					
Czech			5 636	7 378	8 992	9 753	11 794
Slovakia			1 417	2 239	3 091	3 419	3 996
Hungary	3 344	4 915	4 878	6 060	7 611	8 811	11 779
Poland	4 656	7 976	8 458	10 127	12 251	12 245	14 677
Romania	2 664	1 509	1 796	2 795	3 390	3 587	4 537
Total CMEA-5	14 482	21 876	23 199	30 018	37 171	39 516	48 890
CMEA-5 as %							
extra-EU trade	3.37	4.73	4.93	5.79	6.82	6.81	6.91
Slovenia		1 586	3 178	3 798	4 245	4 269	4 715
Estonia		262	296	594	889	1 087	1 657
Latvia		542	727	955	1 126	1 115	1 705
Lithuania		590	699	843	970	1 084	1 376
CEE-10 total		24 856	28 099	36 208	44 401	47 071	58 343
CEE-10 as %							
extra-EU trade		5.37	5.98	6.98	8.15	8.12	8.25

Sources and notes: see Table 5.1.

Table 5.3   EU-15 surpluses in trade with CEE-10, 1989–96 (ECU million)

	1989	1992	1993	1994	1995	1996	1997
Bulgaria	1 120	278	474	333	217	−5	−402
Czechoslovakia	−329	984					
Czech			1 452	1 846	2 661	4 213	4 029
Slovakia			166	−45	101	577	777
Hungary	449	370	1 570	2 006	1 117	1 178	1 638
Poland	−62	1 243	2 657	2 189	3 042	7 582	10 509
Romania	−1920	479	717	100	404	850	459
Total CMEA-5	−743	3 354	7 036	6 429	7 542	14 395	17 010
CMEA-5 % gap	0.15%	1.40%	1.48%	1.17%	0.99%	1.83%	2.16%
Slovenia		−204	441	550	932	1 107	1 576
Estonia		72	196	436	460	607	654
Latvia		−291	−312	−236	−186	−8	−189
Lithuania		−224	−154	13	46	367	745
CEE-10		2 707	7 207	7 192	8 794	16 468	19 796
CEE-10 % gap		1.33	1.51	1.31	1.15	2.07	2.35

Sources: see Table 5.1. '% gap' is the difference between the share of CMEA-5 /CEE-10 in extra-EU exports and extra-EU imports.

basic statistics indicate that there has been a rapid expansion of trade between the CEE-10 and the EU-15 since the collapse of communism. Exports from the EU-15 to the CMEA-5 grew 4.8 fold between 1989 and 1997, while EU-15 imports from the CMEA-5 grew 3.4 fold. Exports from the EU-15 to the CEE-10 grew 2.2 fold between 1993 and 1997 and imports by 2.1 fold. Although the EU-15 remain a substantially more important market and source of imports for the CEE-10 than the latter are for the EU, the share of the CMEA-5 in the exports of the EU-15 (excluding trade between the EU-15 themselves) grew from 3.52 per cent in 1989 to 9.07 per cent in 1997.

However, this rapid growth of trade has created problems for the CEE-10. EU-15 exports to the CEE-10 have grown substantially faster than imports from the CEE-10, resulting in growing EU-15 surpluses in visible trade which reached ECU 16,468 million in 1996 and ECU 17,010 million in 1997. All the CEE-10 economies except Bulgaria (which faced a major financial crisis which forced it to cut back its imports from the EU) and Latvia (whose exports included ECU 426 million of exports of refined oil products derived from crude oil imported from Russia in 1996) incurred deficits in their visible trade with the EU-15 in 1996 and 1997. The share of the CMEA-5 in the imports of the EU-15 rose rapidly from 3.37 per cent in 1989 to 6.82 per cent in 1995. This supports the predictions based on gravity models described in Chapter 1 which anticipated that a rapid redirection of CEE exports would take place as the barriers to trade between the two regions were progressively removed. From 1995 to 1997, however, the share of the CMEA-5 in EU imports from outside the EU has remained roughly constant as EU imports from the CEE-10 have grown at the same rate as EU imports in general. While it is dangerous to generalise on such a few observations, this seems to indicate that the initial period of rapid redirection of CEE exports has ended and that further export growth will depend on improvements in the 'competitiveness' of CEE exports. However, the slow down in trade growth did not affect EU exports to the CEE-10 in 1995–7 and the share of the CMEA-5 in extra-EU exports continued to grow from 7.81 per cent in 1995 to 9.07 per cent in 1997. This has resulted in a growing gap between the share of the CMEA-5 economies in extra-EU exports and extra-EU imports which reached 2.16 per cent in 1997. The corresponding figure for the CEE-10 in 1997 was 2.35 per cent (see Table 5.3).

It is also noticeable from Table 5.3 that the five economies that were nominated by the EU commission for the first round of negotiations for entry accounted for ECU 18,406 million of the total deficit incurred by the CEE-10 in 1997, while the economies in the second tier accounted for only ECU 1390 million. This raises the question of whether the larger deficits incurred by the first-tier economies simply represent improved access to international finance which has enabled them to run deficits on current account, or whether they represent a more basic problems arising from the structure of CEE exports.

5.3   Indebtedness and trade between eastern and western Europe

The argument that the visible trade deficits incurred by the CEE-10, and the five first-tier economies in particular, are largely a transitional phenomenon is based on the theory of the debt cycle.[1] In the first stage of the debt cycle, a new debtor economy is normally a relatively poor economy with a small and undeveloped capital stock, low per-capita incomes, and low savings rates. As a result, domestic absorption (domestic consumption and investment) is greater than domestic production. External credits are used to finance the dual gap between domestic savings and investment and between imports and exports. This permits a higher rate of investment than could be sustained from domestic sources alone. Foreign borrowing also permits the import of capital goods which embody a higher level of technology than that which can be produced domestically. The increase in the size and quality of the capital stock, financed by external credits and foreign investment, stimulates long-term growth and exports. Provided that the government does not pursue an over-expansionary macroeconomic policy and prevents excessive domestic consumption, the increase in output will provide returns to external capital invested in the economy and will generate exports which will allow external debt to be serviced.

It can be argued that the CEE-10 economies satisfy several of the conditions described above. The capital stock inherited from communism, was essentially obsolete. This has created a demand for imports of capital goods which could be expected to generate relatively high rates of social and private return. Domestic financial and banking systems were poorly-developed and were incapable of generating the savings required to finance reconstruction. Consequently, the CEE-10's ability to finance deficits in visible trade, including those with the EU, has reflected the availability of capital inflows since 1989. In the early years of the transition to a market economy, these predominantly consisted of multilateral credits from the IMF, the World Bank and the EU together with bilateral credits from western governments. As the transition has progressed, private capital inflows in the form of foreign direct investment, portfolio investment, and bank credits have become a more important source of finance for the imports required for the modernisation of the industrial capital stock and for investment in infrastructure and for the general restructuring of the CEE-10 economies. The extent to which capital inflows have been used to finance a net inflow of physical capital in the CEE economies is examined in section 5.5.

5.4 The commodity structure of extra-EU trade and trade between the EU-15 and the CEE-10

The basic parameters of the commodity structure of EU-15 trade with economies outside the EU and the combined CEE-10 in 1996 are set out in Table 5.4. The first three columns in Table 5.4 show the commodity structure of EU-15 exports to, and imports from all economies outside the EU while columns 4–6 show the structure of EU imports and exports to the combined CEE-10. The final two columns indicate the percentage share of CEE-10 trade in extra-EU exports and imports of each item.[2] At this level of aggregation the commodity structure of EU-15 trade surpluses and deficits in trade with the CEE-10 broadly corresponds to that of extra-EU trade as a whole. The similarity between the commodity structure of extra-EU trade as a whole and EU-15 trade with the CEE-10 suggests that the structure of supply and demand within the EU itself is a major determinant of the commodity structure of trade between the EU-15 and the CEE-10.

The major exceptions where EU surpluses and deficits in trade with the CEE-10 differ from those in extra-EU trade in general, are food and live animals, organic and inorganic chemicals, office telecommunications and equipment and mineral and metal manufactures. Although the agricultural and farm sectors in the majority of the economies of the CEE-10 account for a significantly larger proportion of employment and GDP than in the EU, the EU-15 made a surplus of ECU 1060 million in trade in food and live animals (SITC 0) with the CEE-10 in 1996, compared with a deficit in extra-EU trade of ECU 11,124 million. The EU-15 incurred a deficit of ECU 624 million in its trade with the CEE-10 in trade in mineral and metal manufactured goods (excluding non-ferrous metals), whilst making a surplus of ECU 21.8 billion in trade with the outside world as a whole. Similarly, in trade in organic and inorganic chemicals, the EU made a surplus of ECU 5.4 billion in extra-EU trade as whole, but a deficit of ECU 236 million in trade with the CEE-10. Finally, the EU incurred a deficit in trade in telecommunications and office

Table 5.4 Commodity structure of extra-EU-15 trade and EU-15 trade with CEE-10 in 1996 (ECU million)

	EXTRA-EU TRADE			EU TRADE with CEE-10			Trade with CEE-10 as per cent	
	Exports	Imports	Balance	Exports	Imports	Balance	EU exports	EU imports
TOTAL (0–9)	623 396	579 990	43 406	63 539	47 071	16 468	10.19	8.12
Primary goods (0,1,2,3,4,68)	77 389	176 093	–98 704	7 399	9 246	–1 847	9.56	5.25
food and live animals (0)	30 832	42 046	–11 214	3 380	2 320	1 060		
other agriculture & forestry (1,21–24,29,4)	17 835	19 079	–1 444	914	2 009	–1 095	8.82	7.08
non-fuel minerals (27,28,68)	10 819	25 803	–14 984	1 021	2 268	–1 247	9.43	8.78
fuels and energy (3)	15 335	79 519	–64 184	1 372	2 423	–1 051	8.95	3.05
Manufactures (5,6 ex 68,7,8)	537 916	388 362	149 554	54 580	37 212	17 367	10.15	9.58
Machinery (7)	281 737	187 165	94 572	26 137	13 186	12 950	9.28	7.05
industrial equipment (70–4)	116 600	47 903	69 825	10 394	4 403	5 991	8.91	9.19
office, telecoms (75–6)	36 555	56 610	–23 055	3 494	1 227	2 267	9.56	2.17
electrical (77)	42 393	40 769	1 624	5 089	3 558	1 531	12.00	8.73
transport equipment (78–9)	86 189	41 883	44 306	7 160	3 999	3 161	8.3	9.54
Chemicals (5)	80 665	44 939	35 726	7 579	3 018	4 561	9.40	6.72
organic and inorganic (51,52)	21 806	16 358	5 448	927	1 163	–236	4.25	7.11
pharmaceuticals (54)	18 712	9 588	9 124	1 418	140	1 278	7.58	1.46
perfumes (55)	8 064	2 185	5 879	915	59	856	11.35	2.73
fertilisers (56)	1 072	1 919	–847	72	632	–560	6.67	32.92
plastics (57–8)	13 933	7 360	6 236	2 142	763	1 379	15.37	10.36
Light consumer goods (81–5)	30 361	48 761	–18 400	3 395	9 650	–6 255	11.18	19.80
Yarns, cotton, leather (61,65)	21 054	15 363	5 691	5 235	1 665	3 570	24.86	10.83
Rubber, cork and paper (62,63,64)	18 767	12 686	6 081	2 665	2 331	334	14.20	18.37
Mineral and metal goods (66,67,69)	55 816	33 941	21 875	5 131	5 755	–624	9.19	16.96

Notes: SITC numbers are given in brackets. Negative sign indicates EU trade deficit.
Sources: COMEXT database and External and Intra-European Union Trade, Eurostat, 1997.

equipment of ECU 23.5 billion in extra-EU trade, but made a surplus in its trade with the CEE-10 of ECU 2.3 billion.

The EU-15 have incurred consistent and substantial deficits in extra-EU trade in all categories of primary goods which have been broadly offset by surpluses in trade in manufactured goods since 1990. The EU deficit in extra-EU trade in primary goods reached ECU 98.7 billion in 1996 of which the CEE-10 accounted for only ECU 2.4 billion. This reflects the fact that the majority of CEE-10 economies are relatively unfavourably endowed with energy resources when compared to major resource-exporting economies. The EU was a major net importer of fuel and energy products with a deficit of ECU 64.2 billion in total, of which only ECU 1.6 billion resulted from trade with the CEE-10. EU-15 deficits largely arose from imports of coal and coke of ECU 1.1 billion which are largely explained by imports from Poland and the Czech Republic of ECU 739 million and ECU 296 million respectively. The EU also incurred a small deficit in trade in refined oil products with CEE-10 exports (which are largely derived from imported Russian crude oil) of ECU 1038 million while EU exports of refined oil products to the CEE-10 only came to ECU 893 million. The EU incurred deficits in trade in non-fuel minerals with the CEE-10 of ECU 1247 million, the largest component of which was a deficit of ECU 810 million in non-ferrous metals. The EU made a small deficit in its trade in agricultural and forestry products with the CEE-10 of ECU 35 million (compared with a deficit of ECU 12.5 billion in extra-EU trade). An EU surplus in trade in food and live animals of ECU 1060 million was offset by an EU deficit of ECU 1143 million in wood and lumber products with the CEE-10.

The EU-15 made substantial surpluses in its trade in manufactured goods with the rest of the world (ECU 149.5 billion) and with the CEE-10 (ECU 17.4 billion). The CEE-10 accounted for 10.15 per cent and 9.58 per cent of extra-EU exports and imports of manufactured goods respectively. EU surpluses in total extra-EU trade in manufactures were largely explained by surpluses in machinery and equipment of ECU 94.6 billion, chemical products of ECU 35.8 billion, mineral and metal products of ECU 21.8 billion, rubber, cork and paper products of ECU 6.1 billion and yarns, cotton and leather of ECU 5.7 billion, while the EU-15 incurred a deficit of ECU 18.4 billion in extra-EU trade in light consumer goods (clothing, footwear, furniture and bedding, travel goods and bags, sanitary and heating and lighting equipment). This pattern was largely reflected in the structure of EU trade with the CEE-10. The EU made surpluses of ECU 12,950 million in trade in machinery and ECU 4561 million in trade in chemicals with the CEE-10 and a deficit of ECU 6255 million in trade in light consumer goods (clothing, furniture, footwear, upholstery, bedding and travel goods).

5.5 The commodity structure of EU-15 trade with individual CEE-10 economies

The commodity structure of EU-15 imports from the individual CEE-10 economies in 1996 is shown in Table 5.5. Table 5.6 shows the commodity structure of EU-15 trade balances with individual CEE-10 economies. A negative sign indicates an EU trade deficit for the given category. The EU was net importer of primary minerals and energy products from each individual CEE-10 economy, but a net importer of food and beverages from only Hungary and Bulgaria. The EU made substantial trade surpluses in manufactured goods in trade with each individual CEE-10 economy. EU surpluses in machinery and equipment were substantial in trade with each CEE-10 economy, reflecting the continuation of trade patterns inherited from the communist era and the demand for machinery and equipment for industrial modernisation and restructuring. The structure of EU trade deficits with individual economies also reflected patterns established in the communist era. The EU was also a net importer of light industrial consumer goods from each individual CEE-10 economy and a net importer of chemical fertilisers from each country except Slovenia. The Czech Republic, Slovakia, Bulgaria and Romania were also net exporters of iron, steel and metal products and non-metallic manufactured goods.

One surprising observation from Table 5.6 is that the EU surplus in trade with machinery and equipment with each of the potential first-tier entrants for accession is smaller than the aggregate surplus with each country. At the same time the EU surplus in trade in machinery and equipment with each of the second-tier economies is larger than the surplus in total trade. This partly reflects differences in the structure of exports from the first and second tier economies and the greater success of the first-tier economies in penetrating EU markets for machinery and equipment. However it may also indicate that the second-tier entrants have been less successful in attracting inflows of foreign capital and have been required to finance imports of machinery and equipment through exports of consumer goods and iron and steel products.

EU surpluses have been broken up into three broad categories which are shown in Table 5.7. These consist of trade in capital goods (the majority of items of machinery and equipment excluding cars and

Table 5.5 Commodity structure of EU-15 imports from individual CEE economies in 1996 (ECU million)

	Poland	Czech	Hungary	Slovakia	Slovenia	Romania	Bulgaria	Estonia	Latvia	Lithuania
Primary goods (0,1,2,3,4,68)	2 873	1 383	1 766	453	323	380	482	420	776	391
Agric & forestry (0,1,21–4,29,4)	1 206	619	1 137	147	130	168	262	193	271	194
Food & beverages (0,1)	893	235	856	50	58	120	193	35	17	62
Minerals and energy (27,28,3,68)	1 621	681	612	286	177	208	203	223	486	193
Manufactures (5–8 excluding 68)	9 271	8 331	7 018	2 951	3 935	3 193	1 207	665	330	690
Machinery (7)	2 770	3 219	3 700	1 023	1 591	406	162	174	29	112
Industrial equipment (70–4)	684	1 274	1 349	273	448	388	111	33	10	10
Office, telecoms, electrical (75–7)	915	945	1 824	262	478	101	36	126	15	82
Transport equipment (78–9)	1 171	1 000	527	488	665	83	15	15	4	20
Chemicals (5)	654	699	526	277	175	194	243	52	31	168
Organic and inorganic (51,52)	292	266	241	71	80	62	93	25	15	15
Pharmaceuticals and perfumes (54)	18	46	29	9	10	12	15	1	0	0
Chemical fertilisers (56)	169	69	12	61	0	68	97	21	6	129
Plastics (57–8)	102	225	196	107	51	46	30	2	1	4
Wood and wood manufactures (63)	452	179	96	40	156	40	19	46	59	28
Iron, steel and metal products (67,69)	1 103	1 213	518	484	336	406	209	52	10	13
Non-metallic minerals (66)	387	475	128	118	110	96	36	16	6	19
Light consumer goods (81–5)	2 854	1 128	1 345	584	866	1 866	416	215	123	253
Furniture (82)	966	373	163	111	280	301	20	53	17	22
Clothing (84)	1 633	486	830	332	439	1 163	297	128	103	218
Footwear (85)	141	118	238	110	74	374	80	21	2	9
Total (0–9)	12 245	9 753	8 811	3 419	4 269	3 587	1 701	1 087	1 115	1 084

Source: estimated from COMEXT database.

108

Table 5.6 Commodity structure of EU-15 balances in trade with individual CEE economies in 1996 (ECU million)

	Poland	Czech	Hungary	Slovakia	Slovenia	Romania	Bulgaria	Estonia	Latvia	Lithuania
Primary goods (0,1,2,3,4,68)	-328	106	-1 040	-40	332	178	-280	-81	-518	-178
Agric & forestry (0,1,21-4,29,4)	281	277	-730	71	209	138	-148	3	-115	-21
Food & beverages (0,1)	346	552	-534	136	208	170	-87	137	112	84
Minerals and energy (27,28,3,68)	-812	-229	-378	-125	61	-6	-152	-90	-390	-174
Manufactures (5-8 excluding 68)	7 645	3 988	2 167	573	640	632	211	632	405	474
Machinery (7)	5 165	3 093	865	877	472	1 037	374	406	278	382
Industrial equipment (70-4)	2 529	1 199	368	117	284	683	99	117	90	162
Office, telecoms (75-76)	739	652	132	217	137	154	63	58	50	64
Electrical equipment (77)	523	689	-13	127	-137	116	68	96	43	15
Transport equipment (78-9)	1 373	552	377	75	187	84	144	135	95	140
Chemicals (5)	2 001	952	649	168	401	247	-22	88	70	6
Pharmaceuticals (54)	504	265	209	81	62	62	20	-23	-8	-7
Perfumes (55)	295	169	121	30	64	61	31	45	31	29
Chemical fertilisers (56)	-143	-47	-6	-59	7	-67	-97	-18	-4	-127
Plastics (57-8)	709	274	98	15	125	37	18	36	15	54
Wood and wood manufactures (63)	-375	-108	-45	-18	-117	-21	-8	-33	-55	-17
Paper and paper board (64)	504	205	292	-15	-2	56	53	45	46	38
Iron, steel and metal products (67,69)	32	-203	82	-261	53	-177	-145	60	16	53
Non-metallic minerals (66)	41	-196	41	-49	9	-29	-5	18	11	3
Yarns, cotton, fibres (65)	1 261	137	458	122	234	636	121	34	29	60
Light consumer goods (81-5)	-2 071	-465	-752	-403	-482	-1 494	-270	-87	-53	-175
Clothing (84)	-1 296	-233	-540	-269	-297	-980	-225	-73	-72	-178
Furniture (82)	-821	-190	-55	-59	-152	-263	2	-25	-1	-6
Footwear (851)	-30	-26	-129	-74	-11	-255	-42	3	8	0
Total (0-9)	7 582	4 213	1 178	577	1 107	850	-5	607	-8	367

Sources and notes: estimated from COMEXT database. A negative sign denotes an EU trade deficit in the given category.

Table 5.7   EU surpluses in trade with CEE-10 in capital goods, consumer goods and intermediate goods (ECU million)

	EU-15 exports of capital goods	EU-15 imports of capital goods	EU surplus in capital goods	EU surplus in consumer goods	EU surplus in other goods
First tier					
Poland	6 189	1 824	4 365	2 025	1 192
Czech	5 299	2 678	2 621	1 635	−43
Hungary	3 748	2 939	809	21	348
Slovenia	1 548	878	670	−11	448
Estonia	427	144	283	339	−15
total	**17 211**	**8 463**	**8 748**	**4 009**	**1 930**
Second tier					
Romania	1 334	385	948	−173	75
Bulgaria	443	165	277	−22	−260
Slovakia	1 592	690	902	82	−407
Latvia	243	27	216	261	−485
Lithuania	414	53	361	170	−164
total	**4 026**	**1 320**	**2 704**	**318**	**−1 241**
CEE-10 total	**21 237**	**9 783**	**11 452**	**4 327**	**689**

Notes: Capital goods = SITC 7 excluding SITC 76, 775, 776, 781, 785 plus SITC 87, 88 (ex 885). Consumer goods = SITC0, 1, 4, 54, 55, 611, 612, 613, 65, 665, 696, 697, 76, 775, 781, 785, 812, 813, 821, 83, 84, 85, 885, 892, 894, 896, 897, 898, 899.

other goods for household consumption plus professional instruments and associated apparatus); items for personal consumption (including foodstuffs, clothing and textiles, textile yarns, furniture, footwear, domestic appliances and goods for household consumption, medicines, perfumes, glassware, jewellery and toiletries) and a residual item that largely consists of energy, raw materials and intermediate goods for industrial processing.[3] Firstly, it is apparent that the major component of the EU surplus in trade with the CEE-10 of ECU 16,468 million in 1996 consisted of a surplus in trade in capital goods of ECU 11,452 million. However the EU also made a surplus in trade with the combined CEE-10 in goods for personal consumption of ECU 4327 million and in materials and intermediate goods of ECU 689 million. The EU surplus in trade in capital goods was the largest component of the aggregate deficit in its trade with each individual CEE-10 economy except Estonia. The pattern for trade in machinery and equipment was repeated in trade in capital goods as a whole, with the EU running smaller surpluses in trade in capital goods with potential first-tier entrants than in total trade, and larger surpluses in trade in capital goods with second-tier entrants than in total trade, with the exception of Lithuania. This indicates that net inflows of financial capital to the more developed economies included in negotiations for early entry to the EU have been used to finance net inflows of items for personal consumption and material inputs for current production in addition to capital goods which will enhance the long-term productive potential of the economy. This is clearly the case for Poland and Estonia, where EU trade surpluses actually exceeded the value of EU *exports* of capital goods. The smaller EU surpluses in trade with Hungary and Slovenia in capital goods largely reflected those countries relative success in penetrating EU markets for capital goods. Hungary has been particularly successful in penetrating EU markets for machinery with exports of ECU 3700 million in 1996 which accounted for 42 per cent of Hungarian exports to the EU (compared with an average of 28 per cent for the CEE as a whole) and for 52.7 per cent of Hungarian exports of manufactured goods. Exports of office equipment, telecommunications and electrical equipment accounted for 20.7 per cent of Hungarian exports to the EU in 1996. Although Hungary had a relatively small deficit in trade in goods for personal consumption (ECU 438 million) this was explained by net exports of food and beverages (ECU 534 million) while trade in industrial consumer goods was in deficit. Slovenia also had a relatively high proportion of exports of machinery and equipment which accounted for 37.2 per cent of total exports and for 40.4 per cent of exports of manufactured goods.

The largest deficits in recorded trade with the EU in 1996 were incurred by Poland (ECU 7582 million) and the Czech Republic (ECU 4213 million) both of whom were substantial net importers of items for personal consumption (ECU 2025 million and ECU 1635 million respectively) including food and beverages, cars, domestic appliances, pharmaceuticals and perfumes. The balance of EU trade with Poland recorded in EU statistics may be misleading in view of the high volumes of 'shuttle trade' with the EU in which unrecorded exports of Polish consumer goods are significant. Despite being a major net exporter of minerals, including coal, coke and non-ferrous metals, Poland was a net importer of materials and intermediate goods of ECU 1192 million largely as a result of high levels of imports of plastic and plastic goods, paper and paperboard, printed matter and office stationery. Consequently, only Hungary, of the economies being considered for entry in the first tier of enlargement could attribute its trade deficit with the EU largely to the inflow of capital required for economic regeneration, while the overall size of the EU surpluses in trade with Poland, the Czech republic and Estonia include substantial surpluses in trade in goods for personal consumption. Populations which had experienced decades of forced savings and had been isolated from western consumer goods have been reluctant to forego the appeal of the newly-available goods carrying western brand names while the CEE economies have been less successful in penetrating EU markets for consumer goods.

The economies in the second-tier of negotiations have financed imports of capital goods through a combination of credits and net exports of consumer goods, raw materials, and intermediate goods. Capital goods accounted for only 12.3 per cent of the exports of the second-tier economies to the EU, compared with 23.4 per cent for the first-tier economies. This, combined with their relative lack of success in attracting capital inflows to bridge the gap has required them to run surpluses in trade in other commodities to finance industrial modernisation. All the second-tier economies, except Romania were net exporters of raw materials and intermediate goods while Romania and Bulgaria were net exporters of items for personal consumption. Romania has run high surpluses in trade in clothing, furniture and footwear but these were insufficient to cover deficits in trade in agricultural products and chemicals as well as machinery. Bulgaria's surpluses in agricultural goods, chemicals and clothing contributed to an overall trade surplus, after years of sustained deficits in the 1980s and 1990s. Latvian exports of minerals and clothing contributed to an overall trade surplus which was slightly lower than in previous years. Lithuania also made a surplus in trade in minerals and clothing which was just insufficient to offset the deficit in machinery, chemicals, iron and steel and paper board.

5.6 The relative importance of the commodity composition of exports for individual economies

It was argued in Chapter 2 that the removal of barriers to trade and the development of the single market in the EU will result in a greater concentration of production structures within the EU. This process will affect the CEE-10 economies as they become integrated into the EU with the effect that patterns of specialisation that were developed under CMEA trading relations may be disbanded altogether, or may be subject to greater rationalisation and concentration of production in fewer locations. There is also the possibility that the first economies to gain accession to the EU will benefit from 'first mover' advantages and will be able to attract capital inflows into their regions at the expense of later entrants, thereby complicating the process of economic recovery in the less-developed regions of Europe. This section will assess the broad areas in which the CEE economies have already demonstrated an export specialisation in relation to one another, to provide an indication of the areas in which they are most likely to benefit from concentration effects. A relatively high level of aggregation will be maintained in this Chapter on the grounds that potential new investors may be more interested in broad areas of specialisation rather than relatively narrow ones (for example, road vehicles in general rather than passenger cars). These estimates will be supplemented by more detailed estimates of the industrial sectors in which the CEE economies are most capable of withstanding competition from existing producers inside and outside the EU in Chapter 7.

Export specialisation indices (ESIs), which were described in Chapter 3, which have been estimated for each of the CEE-10 economies relative to EU imports from the combined CEE-10 are shown in Table 5.8. The size of the ESI indicates the degree to which the share of the given export category for the country in question varies from the share of the category in EU imports from the CEE-10 as a whole, with an index of greater than one indicating an export specialisation. All three Baltic states recorded high ESIs in primary goods which can be attributed to high ESIs for minerals and energy products. Bulgaria, Poland and Hungary also recorded ESIs of greater than one in primary goods, with above-average levels of exports of agricultural products including food and beverages. Although this provides an indication that these countries would benefit from increased investment and concentration of production in the food and beverages sector, attempts to expand this sector at the expense of existing EU producers is expected to encounter major political obstacles. Romania recorded a surprisingly low ESI in primary goods (0.539) which reflect its comparative failure to penetrate EU markets for agriculture and forestry products. These accounted for a smaller share of Romanian exports to the EU than for any other CEE-10 economy, except Slovakia. This can be largely attributed to the neglect of investment in agriculture in the Ceausescu era and the slow pace of agricultural reform since 1989.

The Czech Republic, Slovenia, Slovakia and Romania recorded ESIs of greater than one in manufactured goods. However, the commodity structure of exports of manufactured goods differed substantially from country to country. Exports of machinery and equipment were more important for the Central European economies where Hungary (1.499), Slovenia (1.330), the Czech Republic (1.178) and Slovakia (1.068)

Table 5.8 Export specialisation indices for individual CEE economies in 1996 in relation to total CEE-10 exports to EU

	Poland	Czech	Hungary	Slovakia	Slovenia	Romania	Bulgaria	Estonia	Latvia	Lithuania
Primary goods (0,1,2,3,4,68)	**1.194**	0.722	1.020	0.674	0.385	0.539	**1.442**	**1.967**	**3.543**	**1.836**
Agric & forestry (0,1,21–4,29,4)	**1.071**	0.690	**1.404**	0.468	0.331	0.510	**1.676**	**1.931**	**2.644**	**1.947**
Food & beverages (0,1)	1.363	0.450	1.815	0.273	0.254	0.625	2.120	0.602	0.285	1.069
Minerals and energy (27,28,3,68)	1.329	0.701	0.697	0.840	0.416	0.582	1.198	2.059	4.375	1.787
Manufactures (5–8 excluding 68)	**0.948**	**1.070**	**0.997**	**1.081**	**1.154**	**1.115**	**0.889**	**0.766**	**0371**	**0.797**
Machinery (7)	**0.808**	**1.178**	**1.500**	**1.068**	**1.330**	**0.404**	**0.340**	**0.571**	**0.093**	**0.369**
Industrial equipment (70–4)	0.574	1.343	1.574	0.821	1.079	1.112	0.671	0.312	0.092	0.095
Office, telecoms, electrical (75–7)	0.735	0.953	2.0369	0.754	1.102	0.277	0.208	1.141	0.132	0.744
Transport equipment (78–9)	1.129	1.210	0.706	1.685	1.839	0.273	0.104	0.163	0.042	0.218
Chemicals (5)	**0.833**	**1.117**	**0.931**	**1.263**	**0.639**	**0.843**	**2.227**	**0.746**	**0.433**	**2.416**
Organic and inorganic (51,52)	0.968	1.107	1.110	0.843	0.760	0.701	2.219	0.933	0.546	0.562
Pharmaceuticals and perfumes (54)	0.494	1.586	1.107	0.885	0.788	1.125	2.965	0.309	0	0
Chemical fertilisers (56)	1.028	0.527	0.101	1.329	0	1.412	4.247	1.439	0.401	8.863
Plastics (57–8)	0.513	1.421	1.371	1.928	0.736	0.790	1.087	0.113	0.055	0.227
Wood and wood manufactures (63)	1.558	0.775	0.460	0.494	1.543	0.471	0.472	1.787	2.223	1.091
Iron, steel and metal products (67,69)	0.976	1.348	0.637	1.534	0.853	1.226	1.331	0.518	0.097	0.130
Non-metallic minerals (66)	1.069	1.648	0.492	1.168	0.872	0.906	0.716	0.498	0.182	0.593
Light consumer goods (81–5)	**1.137**	**0.564**	**0.745**	**0.833**	**0.990**	**2.538**	**1.193**	**0.965**	**0.538**	**1.139**
Furniture (82)	1.610	0.781	0.378	0.663	1.339	1.713	0.240	0.995	0.311	0.414
Clothing (84)	1.115	0.417	0.788	0.812	0.860	2.711	1.460	0.985	0.772	1.682
Footwear (85)	0.464	0.488	1.090	1.300	0.699	4.206	1.897	0.779	0.072	0.335

Source: estimated from COMEXT database.

recorded ESIs of greater than unity and were far less important for the Baltic States, Romania and Bulgaria. Within the machinery and equipment category, exports of office equipment, telecommunication and electrical appliances were more important for Hungary (2.037), Estonia (1.141) and Slovenia (1.102) but were relatively unimportant for Romania (0.277) and Bulgaria (0.208). The poor relative performance of Bulgaria in this sector is disappointing in view of the importance of Bulgarian exports of office equipment and data processing equipment to the Soviet Union in the communist era. Similarly, exports of transport equipment were relatively more important for the central-east European economies (except Hungary) but were relatively unimportant for the Baltic States, Romania and Bulgaria.

Exports of chemicals were highly significant for Lithuania (2.416) and Bulgaria (2.227). Bulgaria recorded high ESIs in all sectors of chemicals including fertilisers (4.247), pharmaceuticals and perfumes (2.964) and organic and inorganic chemicals, indicating the relative import-ance of this sector for Bulgarian prospects for integration into EU markets. The share of exports of the traditional heavy industrial products of iron and steel products were most important for Slovakia (1.534), the Czech Republic (1.348), Bulgaria (1.33) and Romania (1.23) of whom, only the Czech Republic, is a candidate for the first round of enlargement.

Exports of light consumer goods (Sitc 81–5) were highly significant for Romania (2.538) and marginally important for Bulgaria (1.192), Lithu-ania (1.138) and Poland (1.137). Romania had the highest ESI in each of furniture (1.712), clothing (2.711) and footwear (4.206). Exports of fur-niture (which includes fabrics, cushions etc.) were also relatively import-ant for Poland and Slovenia. Exports of clothing were relatively important for Bulgaria, Lithuania and Poland and footwear for Bulgaria, Slovakia and Hungary.

As far as manufacturing industry is concerned, existing patterns of specialisation indicate that production of machinery and engineering products should be concentrated on Hungary, Slovenia, the Czech Republic and Slovakia, with Hungary and Slovenia having a specialisa-tion in telecommunications and more sophisticated electrical products. Bulgaria, Lithuania and the Czech Republic reveal a specialisation in the production of chemicals, although in Lithuania this is confined to fertil-iser production, while all other categories of chemicals fell below the average weighting of the CEE-10 economies. Bulgaria and the Czech Republic display a higher level of specialisation in more sophisticated chemical products including pharmaceutical goods and perfumes. The

Czech Republic, Slovakia, Bulgaria and Romania have a specialisation in iron and steel products, although in the case of Bulgaria and Romania in particular this may result from investment decisions in the communist era that were not justified by market conditions. Romania, Poland and Bulgaria have a specialisation in light industrial consumer goods (furniture, footwear and clothing). However, exports of these goods are heavily dependent on outward processing trade which will be examined in the next section.

5.7 Outward processing trade between the EU-15 and the CEE-10

5.7.1 The structure of EU outward processing trade with the rest of the world and the CEE-10

It can be seen from Table 2.4 that trade between the EU-15 and the CEE-10 in textile yarns and light consumer goods is more intense than in extra-EU trade in general. The CEE-10 accounted for 24.9 per cent of the EU-15 exports of yarns, fabrics and leather goods compared with a share of 10.15 per cent of EU exports of manufactured goods as a whole. Similarly, the CEE accounted for 19.8 per cent of EU imports of clothing, furniture and footwear, compared with a share of 9.58 per cent of EU imports of all manufactured goods. The higher proportion of bilateral trade in textile yarns and clothing reflects the greater share of trade based on outward processing agreements (in which trade in clothing and textiles predominates) in economic relations between the EU and the CEE-10, than in extra-EU trade in general. Outward processing trade (OPT) can be seen as a specialised form of international subcontracting in which the principal (the contractor) supplies the agent (the subcontractor) with the materials for processing which are then delivered back to the principal or to the customer of the principal.[4] When this form of trade takes place between enterprises based in different countries inside the EU, there is no need for formal registration of the arrangement and trade flows in both directions are simply recorded as normal exports and imports in foreign trade statistics. However, when such trade takes place between enterprises inside and outside the EU, the inputs which have been exported to the subcontractor for processing and re-import must be registered with the EU customs authorities to avoid the payment of EU tariffs on the re-imported materials. Consequently a statistical record exists for such trade between EU and non-EU partners, although it is not available for trade between EU partners.

Table 5.9   EU Outward processing trade in manufactured goods with world and CEE-10 in 1996 (ECU million)

	EXTRA-EU TRADE			EU TRADE with CEE-10		
	Exports	Imports	Balance	Exports	Imports	Balance
Total (5–8)	13 460	13 575	−116	5 355	6 234	−878
Chemicals (5)	141	284	−143	73	67	6
Furs, leather (61)	240	15	225	172	14	158
Manufactured materials (6)	4 407	572	3 835	3 020	433	2 587
Textile yarns, fibres (65)	3 754	380	3 374	2 586	324	2 262
Minerals and metals (66–69)	288	146	142	202	75	127
Machinery (7)	6 781	5 307	1 474	1 086	911	175
electrical (77)	3 877	2 321	1 556	641	435	206
Light consumer goods (81–5)	1 235	6 778	−5 543	888	4 659	−3 771
Clothing, apparel (84)	1 063	6 180	−5 118	785	4 259	−3 474
Footwear (85)	129	371	−241	70	208	−138
Furniture (82)	28	186	−158	24	173	−149

Source: estimated from COMEXT database.

The major details of EU-15 trade with the CEE-10 and in extra-EU trade in total under OPT arrangements are shown in Table 5.9. OPT is not of major significance in total extra-EU trade and accounted for only 2.5 per cent of EU exports of manufactured goods (including non-ferrous metals) and for 3.4 per cent of EU imports of manufactures in 1996. However OPT is of far greater importance in EU trade with the CEE-10 than in extra-EU trade in general. OPT arrangements accounted for 9.6 per cent of EU-15 exports of manufactured goods and for 16.1 per cent of EU imports of manufactures while imports from the CEE-10 accounted for 45.9 per cent of EU-15 imports under OPT arrangements.

Although EU imports from the CEE-10 contain a larger proportion of goods in which OPT is significant, this is not the sole explanation for the higher proportion of OPT in imports from CEE-10 economies. The weighting of OPT trade in EU imports of specific goods such as clothing, furniture and footwear from the CEE-10 is significantly higher than in total extra-EU imports of these goods. OPT is most important in the clothing and textile industries where it accounted for 45.5 per cent of EU imports of manufactured goods under OPT arrangements from all sources, and for 18.4 per cent of all EU imports of clothing from outside the EU. Clothing accounted for 74.7 per cent of EU imports from the CEE-10 under OPT. EU imports of clothing from the CEE-10 in 1996 under OPT arrangements amounted to ECU 4259 million out of clothing imports of ECU 5470 million. OPT accounted for 77.9 per cent of EU imports of clothing from the CEE-10 compared with only 6.8 per cent of EU imports of clothing from non-EU economies other than the CEE-10. The CEE-10 were the source of 68.7 per cent of EU imports of clothing under OPT arrangements. Just over half (52.7 per cent) of EU exports of textiles yarns and fibres and fabrics and furs to the CEE-10 were exported as part of OPT arrangements compared with 7.4 per cent of exports of yarns and fibres to non-CEE economies.

The other major areas in which OPT arrangements were important for EU trade with the CEE-10 were footwear, where OPT accounted for 17.8 per cent of EU imports of footwear from the CEE-10 (compared with 3.4 per cent of imports from the rest of the world), electrical goods where OPT accounted for 12.2 per cent of EU imports from the CEE-10 (compared with 5.1 per cent of imports from the rest of the world) and furniture where OPT accounted for 6.5 per cent of imports from the CEE-10 but was negligible in EU trade with countries other than the CEE-10.

The greater importance of OPT in EU trade with the CEE-10 may be attributed to three principal factors. Firstly, geographical proximity reduces the cost of exporting materials and components from the EU-15 to the CEE economies for outward processing, compared with more distant low-wage regions where it may be cheaper to use domestic (non-EU) resource-based inputs (for example, wood for furniture). Secondly, OPT arrangements with the EU have partly replaced CEE trade with the former Soviet Union, where the CEE exporter depended on imported materials from the USSR that cannot be obtained from the domestic market. As a result the EU manufacturer has supplied basic inputs from other EU sources. Thirdly, the quality specifications of CEE production of basic inputs (e.g. textile yarns, cotton, treated leather, electrical components) may still fail to meet EU standards and have been replaced by materials and components produced in the EU. Finally, the existence of EU tariffs on imports from CEE economies increases the incentive for manufacturers in EU countries to source materials from other EU suppliers. In the last two cases, it would be expected that OPT would become of reduced importance as the CEE economies improve the quality standards of domestically produced inputs and as tariffs are removed on imports from CEE economies.[5]

5.7.2   The importance and structure of outward processing trade for the individual CEE-10 economies

The commodity structure of trade in manufactured goods between the EU-15 and the individual CEE-10 economies is shown in Table 5.10. The top section indicates the value of EU imports of finished and semi-finished manufactures from CEE-10 economies which have been produced under OPT. The lower section indicates deliveries of materials and components from EU suppliers to individual CEE-10 economies for outward processing. Exports of light consumer goods, especially clothing, predominate in each country's OPT exports, with the exception of Estonia and the Czech Republic. Similarly supplies of yarns and fibres and semi-finished clothing predominate in EU exports to each CEE-10 economy except the Czech Republic and Estonia. Consequently, the weight of OPT in total exports of manufactured goods for each country is closely linked to the importance of clothing in the total exports of the country concerned. Romania, which has the highest share of clothing in exports of manufactured goods to the EU (36.4 per cent) is the most dependent on OPT with 32.9 per cent of exports to the EU produced under OPT. Lithuania, Latvia, Poland and Bulgaria who are less dependent on clothing exports than Romania, have intermediate proportions of trade that is conducted under OPT. However, OPT was not confined to clothing and other light consumer goods such as footwear and furniture and there are growing signs of OPT in machinery and equipment in central Europe. Just under 10 per cent of EU imports of machinery from the Czech Republic were conducted under OPT, including electrical components, car parts, gears and transmission engines while imports of machinery under OPT were also significant for Hungary and Poland.

Table 5.11 shows the effect of netting off exports undertaken under OPT on the levels of CEE-10 exports of light consumer goods. CEE-10 exports of clothing and apparel fall from ECU 5630 million to only ECU 1371 million. Exports of furniture now become more important to the combined CEE-10 economies than exports of clothing. Assuming that a majority of the inputs that are used in the production of furniture are not imported from other regions (and utilise replaceable materials) this may give a more accurate picture of the contribution of domestic value-added to the structure of exports and of the exports to sustaining total employment (including industries supplying inputs to the sector) in the region.

Table 5.10 Outward processing trade in manufactured goods between the EU-15 and individual CEE-10 economies in 1996 (ECU mn)

	Poland	Hungary	Czech	Slovakia	Slovenia	Romania	Bulgaria	Estonia	Latvia	Lithuania
EU Imports										
Total (5–8)	1 886	1 140	1 024	346	243	1 051	240	74	67	161
per cent of imports of manufactures	20.3%	15.9%	12.2%	11.7%	6.2%	32.9%	19.9%	11.1%	20.3%	23.3%
Materials etc. (6)	**146**	**86**	**145**	**20**	**4**	**22**	**1**	**5**	**1**	**3**
Yarns, fibres (65)	131	74	94	9	3	6	0	4	0	3
Minerals, metals (66–9)	11	5	38	2	1	16	1	1	0	0
Machinery (7)	**134**	**268**	**365**	**53**	**12**	**24**	**4**	**45**	**7**	**0**
Electrical (77)	71	132	172	16	4	10	1	22	7	0
Light consumer goods (81–5)	**1 574**	**752**	**388**	**264**	**214**	**995**	**233**	**23**	**59**	**158**
Clothing (84)	1 419	670	346	244	206	926	209	22	59	158
Footwear (85)	24	58	24	9	7	64	21	0	0	0
Furniture (82)	127	18	14	8	1	5	0	0	0	0
EU Exports										
Total (5–8)	**1 400**	939	1 164	317	199	846	212	110	56	111
Materials etc. (6)	**979**	**443**	**388**	**175**	**143**	**618**	**136**	**21**	**35**	**80**
Yarns, fibres (65)	910	355	238	154	133	542	125	17	34	79
Minerals, metals (66–9)	26	19	108	6	6	31	1	3	0	1
Machinery (7)	**137**	**244**	**498**	**86**	**12**	**18**	**2**	**81**	**8**	**0**
Electrical (77)	75	134	317	54	3	7	1	42	7	0
Light consumer goods (81–5)	**208**	**214**	**138**	**43**	**31**	**155**	**62**	**4**	**11**	**24**
Clothing (84)	194	176	119	39	28	139	52	3	11	24
Footwear (85)	5	26	9	3	2	15	9	0	0	0
Furniture (82)	7	11	5	0	1	0	0	0	0	0

Source: estimated from COMEXT database.

Table 5.11   CEE-10 Exports of light consumer goods net of outward processing trade (ECU million)

	Total	Clothing	Footwear	Furniture
Poland	1 280	214	118	839
Hungary	593	160	180	145
Czech	740	141	94	359
Slovakia	324	88	101	103
Slovenia	652	233	67	279
Romania	872	237	310	296
Bulgaria	183	88	59	20
Estonia	192	106	21	53
Latvia	64	44	2	16
Lithuania	95	60	9	22
Total	4 995	1 371	959	2 132

Source: estimated from COMEXT database.

5.8   Unit values of EU imports from the CEE-10

Unit values indices (UVIs) provide basic indicators of the quality levels of EU imports from the CEE-10 (see Chapter 2). The UVI is expressed as the value of EU imports of the good in thousand ECU divided by the weight of imports of the good in metric tonnes. UVIs have been calculated for EU imports from the rest of the world and from the individual CEE economies. Table 5.12 shows the ratios of the UVIs of EU imports from the individual CEE-10 economies to the UVIs of EU imports of the same good from the rest of the world.[6] An index of greater than one indicates that the exporting country is operating in the higher quality segments of the market for the given product, or group of products, while an index of less than one indicates that the exporting country is trading in the lower quality segments of the market.[7]

Table 5.12 provides a comparison of the relative quality levels of the exports of the CEE economies in EU markets. It is immediately noticeable that the CEE economies score UVI ratios of substantially less than one across an exceedingly wide range of products at the two digit level of the SITC classification and that no single economy scores a UVI ratio of greater than one at any single digit level of the SITC product nomenclature. This indicates that the CEE economies, in general, failed to produce the types of goods that sell at relatively high prices in EU markets and that the products that were successfully exported by CEE economies received lower prices than those received by other exporters

Table 5.12 Ratio of unit values of EU imports of manufactured goods from individual CEE economies to unit value of EU imports from the world in 1996

	Poland	Czech	Hungary	Slovakia	Slovenia	Romania	Bulgaria	Estonia	Latvia	Lithuania
All manufactured goods (5–8)	**0.261**	**0.323**	**0.713**	**0.242**	**0.717**	**0.305**	**0.163**	**0.251**	**0.224**	**0.144**
Chemicals (5)	0.321	0.426	0.545	0.320	0.518	0.247	0.176	0.196	0.271	0.163
Organic (51)	0.594	0.562	0.531	0.574	0.281	0.503	0.371	0.441	0.572	2.254
Inorganic (52)	0.478	0.477	0.535	0.565	1.086	0.520	0.301	0.538	0.403	0.462
Pharmaceuticals (54)	0.100	0.330	0.243	0.884	0.576	0.160	0.249	0.047	na	0.178
Fertilisers (56)	0.923	0.803	1.176	0.876	1.300	0.999	0.982	1.159	0.978	1.052
Plastics primary (57)	0.487	0.549	0.576	0.465	0.336	0.452	0.563	0.911	0.857	0.547
Plastics non primary (58)	0.467	0.644	0.387	0.351	0.521	0.351	0.399	0.416	na	0.172
Machinery	**0.231**	**0.284**	**0.490**	**0.308**	**0.368**	**0.192**	**0.171**	**0.434**	**0.241**	**0.179**
Power generating (71)	0.214	0.233	0.521	0.223	0.336	0.141	0.227	0.164	0.289	0.234
General purpose equipment (74)	0.274	0.435	0.576	0.343	0.458	0.342	0.240	0.368	0.278	0.516
Computers, office equipment (75)	0.471	0.262	0.752	0.236	0.581	0.409	0.496	0.434	2.027	3.190
Telecoms and recording (76)	0.404	0.518	0.577	0.908	0.334	0.422	0.354	3.027	0.089	0.235
Road vehicles (78)	0.520	0.535	0.719	0.997	1.083	0.511	0.218	0.347	0.242	0.591
Passenger cars (781)	0.633	0.750	0.998	1.433	1.223	0.708	1.121	0.607	1.061	0.678
Miscellaneous manufactures classified by material (6)	**0.332**	**0.406**	**0.704**	**0.327**	**0.975**	**0.306**	**0.347**	**0.331**	**0.355**	**0.272**
Cork and wood manufactures (63)	0.572	0.586	0.700	1.004	1.629	0.936	0.431	0.479	1.407	0.401
Iron and steel (67)	0.837	0.799	0.771	0.837	1.954	0.788	0.609	0.837	0.547	0.555
Non-ferrous metals (68)	0.886	0.870	0.850	0.681	0.854	0.635	0.705	1.106	1.459	1.627
Miscellaneous manufactures (8)	**0.392**	**0.457**	**0.762**	**0.481**	**0.627**	**0.558**	**0.797**	**0.407**	**0.648**	**0.613**
Furniture (821)	0.557	0.537	0.779	0.468	0.593	0.372	0.322	0.272	0.277	0.217
Clothing (84)	1.452	1.424	1.607	1.362	2.479	1.223	0.920	1.402	1.174	1.061
Footwear (851)	0.972	1.124	1.856	1.546	1.537	1.211	1.076	1.280	0.911	0.789

Source: estimated from COMEXT database.

to the EU.[8] UVIs over, or close to one were concentrated in resource-intensive and labour-intensive goods including fertilisers, non-ferrous metals (especially from the Baltic states) cork and wood manufactures and clothing and footwear. Some isolated positive UVI ratios were scored on products with low volumes of exports (for example cars from Bulgaria, telecommunications and computing equipment from Estonia, Latvia and Lithuania). UVI ratios in traditional export items like iron and steel were less than one for each individual CEE.

Table 5.12 provides some insights into the relative performance of the CEE-10 in EU markets. Slovenia and Hungary both record substantially higher UVI ratios than the other CEE-10 economies in total trade in manufactured goods. In the case of Hungary, in particular, this reflects both greater success in penetrating higher-value markets and higher UVI ratios across a wide range of products. Hungary recorded the highest UVI ratio in three of the four single digit categories. In addition to selling a higher proportion of capital-intensive products embodying high and medium technology in the machinery and chemicals sectors (see Chapter 6), Hungary also recorded relatively high UVIs in labour-intensive goods including clothing, furniture and footwear and recorded the highest UVI in fertilisers. Slovenia recorded positive UVI ratios across a range of products including inorganic chemicals, fertilisers, road vehicles, cork and wood manufactures, iron and steel and clothing and footwear indicating that it can attain EU quality levels in a number of sectors of manufacturing. The performance of the other CEE-10 economies, none of whom (including economies in the first-tier of negotiations) obtained an overall UVI ratio of 0.333 in trade in all manufactured goods, was far less impressive and reflected a general pattern of dependence on exports in low quality segments of EU markets. The ratio of the unit value of manufactured exports from Bulgaria and Lithuania were exceedingly low, reflecting the concentration of exports in low value commodities, while the unit value of aggregate Romanian exports was boosted by the imported component of goods produced for re-export.

5.9   Conclusions

The growth of trade between the EU-15 and the CEE-10 has been significant since 1989 and has helped to compensate the CEE economies for the collapse of the CMEA market. However, EU exports to the CEE-10 have grown substantially faster than EU imports from the CEE-10 resulting in a growing EU trade surplus which reached ECU 19.8 billion in 1997. The share of goods from the CEE-10 in EU imports stagnated in the mid 1970s while the share of the CEE-10 in EU exports continued to grow. The EU surplus can be partly, but not entirely, explained by the CEE demand for capital goods for modernisation and reconstruction, the import of which has been assisted by net capital inflows into the CEE economies from multilateral financial agencies and private sources. The more developed economies which are candidates for accession to the EU in the first tier of enlargement have been more successful in attracting net capital inflows than the economies in the second tier of negotiations, which has allowed the first tier economies to run larger current account deficits. However, capital inflows have also been used to finance net inflows of items for personal consumption and intermediate goods. The economies which are not expected to enter the EU in the first round of enlargement have been less successful in attracting net capital inflows and have financed imports of capital goods partly through net exports of raw materials and intermediate goods and, in the case of Romania, goods for personal consumption.

The enlargement of the EU is expected to result in greater rationalisation of production and concentration of production in a smaller number of CEE economies. The Baltic states were specialised in the export of minerals and energy products and Bulgaria, Poland and Hungary in food and beverages. Hungary, Slovenia, the Czech Republic and Slovakia were specialised in the export of machinery, with Hungary and Slovenia showing a specialisation in the export of more sophisticated equipment, including office machinery and electrical appliances. Bulgaria, Lithuania and the Czech Republic were specialised in chemical exports, with the Czech Republic and Bulgaria showing a specialisation in pharmaceuticals and perfumes. Romania, Poland and Bulgaria were specialised in the export of light consumer goods which were largely produced under outward processing agreements. Outward processing trade is of far greater significance in EU trade relation with the CEE economies than in trade with the rest of the world. Estimates of the relative unit values of EU imports from CEE economies compared with imports from the rest of the world indicate that the CEE economies have failed to penetrate EU markets for high value products and were receiving substantially lower unit values than competitors from outside the EU across a wide range of products.

6
Factor Intensity and Technological Levels in Trade in Manufactured Goods between the CEE-10 and the EU

6.1 Introduction

This Chapter will analyse the technological level and factor-intensity of trade flows in manufactured goods between the CEE-10 and the EU-15, using the Wolfmayr-Schnitzer (1998) adaptation of the Legler–Schulmeister system of classification which was described in Chapter 3. The Chapter will examine the extent to which the product structure of trade between the CEE-10 and the EU is determined by differences in technological levels and factor endowments, and the extent to which trade between the EU-15 and the CEE-10 differs from trade between the existing members of the EU themselves and from trade between the EU and other economies outside the EU. This will help to place the export structure of the CEE economies in a comparative context with those of other EU economies and economies in south-east and central Asia and Latin America. The Chapter will also examine differences between the factor content and technological level of exports and net trade flows of the individual CEE-10 economies to the EU to assess the relative standing of individual transition economies.

Trade in manufactured goods between the EU-15 and the CEE-10 economies and a number of other countries have been broken down into three broad categories in Tables 6.1–6.9. These categories consist of products that are produced by human capital-intensive methods, products produced by labour-intensive methods and products produced by resource-intensive methods (see Chapter 3). Products embodying human capital have been further subdivided into products which require high, medium and (comparatively) low technology processes of production. Table 6.1 shows the factor content of trade in manufactured goods between the EU-15 and the CEE-10 in 1996, expressed in millions of

ECUs. The size of the EU surplus in trade with the CEE-10 in manufactured goods was such that the EU actually made a surplus in trade in each factor-content category and sub-category in trade in manufactured goods except resource-intensive goods, but including labour-intensive goods.

This supports the argument outlined in Chapter 3 that measures of the net factor content, and technological levels, of trade in manufactured goods between the EU and the CEE economies should be based on the share of the respective items in imports and exports, not on absolute volumes of trade. Net factor contents in Tables 6.2, 6.3, 6.5 and 6.8 have been estimated as the difference between the percentage share of a given category in EU imports of manufactured goods from a partner country minus the percentage share of the same category in EU exports of manufactured goods to the partner country. As the purpose of this Chapter is to facilitate comparisons between the structure of trade between the CEE economies and the EU and trade between other economies and the EU, net factor flows in Tables 6.2 onwards have been referred to from the perspective of the EU trade partner country, not the EU itself, for purposes of consistency. Consequently, a positive sign in the net factor content column indicates that the partner country is a net exporter of the category (in relative terms) to the EU. This has the disadvantage that it makes the explanation of Table 6.2 cumbersome. The Chapter concludes by examining the changes in the factor content of exports from the CMEA-5 to the EU following the collapse of communism.

Table 6.1 Factor content of trade in manufactured goods between EU-15 and CEE-10 (ECU million)

	EU-15 exports to CEE-10	EU-15 imports from CEE-10	Balance
HUMAN CAPITAL-INTENSIVE	31 877	14 674	+17 204
High technology	5 103	1 812	+3 291
Labour-intensive	2 546	1 398	+1 148
Captial-intensive	2 557	414	+2 143
Medium technology	22 237	9 550	+12 687
Labour-intensive	13 010	5 341	+7 669
Captial-intensive	9 227	4 209	+5 017
Low technology	4 538	3 312	+1 226
LABOUR-INTENSIVE	16 835	15 810	+1 025
RESOURCE-INTENSIVE	6 365	8 252	−1 887
TOTAL	55 078	38 736	16 342

Source: estimated from COMEXT database.

6.2   Problems created by resource-intensive exports

Resource-intensive exports are important for the CEE economies. Resource-intensive exports accounted for 21.3 per cent of EU imports of manufactured goods from the CEE in 1996 compared with only 12.0 per cent of EU imports from the rest of the world . While EU trade in resource-intensive goods with the rest of the world was roughly balanced, the CEE were substantial net exporters of resource-intensive goods (see Table 6.2). The availability of resource-intensive exports reduces the share of other categories, including products embodying human capital-intensive processes and high technology in total exports and biases downwards estimates of technological levels, based on shares of human capital-intensive goods in total exports of manufactures. The high level of resource-intensive exports could simply reflect an abundant supply of natural resources which it is economic to process and export, even though the country has a high level of endowment of human capital. Should resource-intensive exports be excluded from comparisons of the factor content of manufactured goods in the same way that other primary goods (for example, refined oil products) have been excluded?

There are three strong arguments in favour of including resource-intensive goods in comparisons of the technological level of CEE exports. Firstly, economic development tends to result in the accumulation of physical and human capital which reduces the importance of trade that is determined by pure natural-resource endowment. A high level of education and training also raises the opportunity cost of labour, driving up wage rates and making it less profitable to extract natural resources.[1] Secondly, CEE exports of resource-intensive goods to the EU, largely consist of basic iron and steel products, simple chemical fertilisers and non-ferrous metals. These reflect the pattern of industrialisation inherited from the Soviet period which used relatively simple industrial processes and depended on Soviet supplies of materials, and on some occasions low-quality domestic resources including brown coals and lignite. The continued exports of these goods by some countries in the post-communist era suggests a relative failure to develop alternative sources of exports. This argument is reinforced by the size of current account deficits incurred by the CEEs in the mid 1990s. Consequently information on levels of exports of resource-intensive goods is particularly useful for making comparisons between CEE economies and for examining changes in the export structure of individual CEEs through time.

In theory, the existence of resource-intensive exports should drive up exchange rates and make some other exports uncompetitive. It will be informative to know the extent to which resource-intensive exports displace labour-intensive exports or human capital-intensive exports. If labour is abundant in relation to human capital, some of the burden of adjustment may fall on wage-rates to preserve the competitiveness of labour-intensive exports. This will be reflected in a relatively high level of labour-intensive exports in relation to human capital-intensive exports. Consequently, two additional indices have been introduced in Tables 6.2–6.8 to supplement the basic measurements of factor proportions. The first measures the ratio of trade in human capital-intensive products to labour-intensive products in both exports and imports. The second measures the ratio of the value of high-technology products to labour-intensive products in both exports and imports. Net factor flows have been measured as the ratio of the corresponding ratio of EU imports to EU exports to any given country or region. An index of less than 1 indicates that the partner country has a smaller ratio of human capital or high technology products to labour-intensive products in its exports to the EU than in its imports from the EU while an index of greater than one indicates that the country concerned has a higher ratio of human capital/ high technology goods to labour-intensive goods in its exports than in its imports.

6.3   Factor proportions in EU trade

6.3.1   Basic factor proportions in CEE and world trade with the EU

Table 6.2 provides a summary of the factor proportions embodied in the trade in manufactured goods between the EU-15 and the combined CEE-10, compared with intra-EU trade and the trade of non-EU countries with the EU-15. It is noticeable that goods which embody human capital dominate EU trade in manufactured goods in both intra-EU trade and trade with the rest of the world. Human capital-intensive goods accounted for 60.0 per cent of intra-EU trade in manufactured goods in 1996 and for 65.3 per cent of world imports from the EU and a slightly lower proportion (57.9 per cent) of CEE-10 imports from the EU. High technology goods (which fall within the category of human capital-intensive goods) accounted of 14.7 per cent of intra-EU trade and for 16.8 per cent of world imports from the EU but for only 9.3 per cent of CEE-10 imports from the EU, reflecting a lower absorptive capacity for high-technology products in transition economies. Labour-intensive

Table 6.2 Factor content of EU-15 trade in manufactured goods in 1996 (percentages of exports of manufactures)

	Intra-EU trade	World exports to EU	World imports from EU	Net factor content world-EU	CEE-10 exports to EU	CEE-10 imports from EU	Net factor content CEE-EU
HUMAN CAPITAL-INTENSIVE	60.0	58.5	65.3	–6.8	37.9	57.9	–20.0
High technology	14.7	22.8	16.8	+6.0	4.7	9.3	–4.6
Labour-intensive	7.6	14.5	11.0	+3.5	3.6	4.6	–1.0
Capital-intensive	7.1	8.3	5.8	+2.5	1.1	4.7	–3.6
Medium Technology	37.8	29.2	40.3	–11.1	24.7	40.4	–15.7
Labour-intensive	17.0	19.4	23.8	–4.4	13.8	23.6	–9.8
Capital-intensive	20.8	9.8	16.4	–6.6	10.9	16.8	–5.9
Low technology	7.5	6.5	8.2	–2.7	8.5	8.2	+0.3
LABOUR-INTENSIVE	24.7	29.5	23.5	+6.0	40.8	30.6	+10.2
RESOURCE-INTENSIVE	15.3	12.0	11.2	+0.8	21.3	11.5	+9.8
Human capital:labour-intensive ratio	2.43	1.98	2.78	0.71	0.93	1.89	0.49
High tech:labour-intensive ratio	0.60	0.77	0.71	1.08	0.11	0.30	0.37

Source: estimated from COMEXT database. Methodology derived from Wolfmayr-Schnitzer (1998).

goods accounted for 24.7 per cent of intra-EU trade and for 23.5 per cent of world imports from the EU. As a result the ratio of human capital-intensive products to labour-intensive products in intra-EU trade came to 2.43 and high-technology products to labour-intensive products to 0.60 and to 2.78 and 0.71 in world imports from the EU.

Human capital-intensive goods also predominated in world exports to the EU accounting for 58.5 per cent of EU-15 imports of manufactured goods with high-technology goods accounting for 22.8 per cent, labour-intensive goods for 29.5 per cent and resource-intensive goods for 12 per cent. The rest of the world are revealed as net importers of human capital-intensive goods from the EU (−6.8 per cent) but net exporters of manufactured goods which embody high technology (+6.0 per cent). The counterpart of the above is that the rest of the world were net exporters of labour-intensive goods (+6.0 per cent) and resource-intensive goods (+0.8 per cent).

The structure of imports of the combined CEE-10 from the EU-15 broadly resembles the structure of intra-EU trade. Human capital-intensive goods account for 57.9 per cent of CEE-10 imports compared with 60.0 per cent of intra-EU trade. CEE-10 imports of labour-intensive goods are higher at 30.6 per cent than in intra-EU trade at 24.7 per cent, largely as a result of CEE-10 imports of labour-intensive goods which are used in outward-processing arrangements before re-export. The level of CEE-10 imports of resource-intensive exports is correspondingly lower at 11.5 per cent compared with 15.3 per cent.

However, there are major divergences between the structure of CEE-10 exports to the EU and intra-EU trade which result in trade flows based on differences in factor content between the two regions. Human capital-intensive goods accounted for only 37.9 per cent of CEE-10 exports of manufactured goods to the EU-15 and goods embodying high-technology for only 4.7 per cent, less than a third of the proportion in intra-EU trade and just over a fifth of the proportion of world exports of high-technology goods to the EU. Labour-intensive goods accounted for 40.8 per cent of CEE-10 exports and resource-intensive goods for 21.3 per cent. As a result the CEE-10 were substantial net-importers of goods embodying human capital from the EU (−20.0 per cent) and net importers of high-technology goods (−4.6 per cent). The CEE-10 were net exporters of labour-intensive goods (+10.2 per cent) and resource-intensive goods (+9.8 per cent) in trade in manufactures with the CEE-10. The ratio of CEE-10 exports of human capital-intensive goods to labour-intensive goods was 0.93 compared to a ratio of 2.43 in intra-EU trade and EU and 1.89 in EU exports to the CEE-10. This discrepancy is more marked in exports of high-technology goods where the ratio to labour-intensive goods stood at 0.11 for CEE-10 exports to the EU compared with 0.60 in intra-EU trade and 0.30 for CEE-10 imports from the EU.

6.3.2 A comparison of CEE trade with intra-EU trade

How do the trade flow of the CEE-10 economies as a whole compare with those of individual members inside the EU? Table 6.3 provides comparative data on the factor content of trade flows of three poorer members of the EU (Greece, Portugal and Spain) with the remainder of the EU. Major differences can be observed between the structure of exports from these three countries themselves to the remaining members of the EU. The structure of Spain's trade with the remainder of the EU does not differ significantly from that of intra-EU trade as a whole shown in Table 6.1. Human capital-intensive goods accounted for 60.2 per cent of Spain's exports to the EU and for 60.6 per cent of Spain's imports from the EU, labour-intensive goods for 25.4 per cent of Spain's exports and 25.6 per cent of Spain's import and resource-intensive goods for 14.4 per cent and 14.6 per cent respectively. The ratio of human capital-intensive products to labour-intensive products was 2.37 for both imports and exports compared with 2.43 for intra-EU trade as a whole. As a result Spanish trade with the remainder of the EU could not be explained by differences in the factor content of imports and exports which were relatively small in each category. The only significant imbalance fell *within* the category of human capital-intensive goods where Spanish deficits in trade in high technology goods (−4.9 per cent) and low-technology goods (−3.8 per cent) were offset by surpluses in medium-technology products (+8.3 per cent). The structure of both Portugal's and Greece's trade with the EU was considerably less advanced than that of Spain with labour-intensive goods accounting for just over 50 per cent of both countries exports to the remainder of the EU. However, there was a marked disparity between the proportion of exports of human capital-intensive goods from Portugal, which accounted for 38.3 per cent of total exports, and from Greece where they accounted for only 19.5 per cent of exports. As a result Greece had a deficit (net inflow) in trade in human capital-intensive goods of 37.9 per cent compared with a deficit of 16.6 per cent for Portugal. The ratio of human capital-intensive to labour-intensive goods in Greek exports was 0.38 and 0.76 in Portuguese exports compared with 0.93 for the CEE-10.

Consequently, the structure of exports of the combined CEE-10 to the EU was technologically inferior to that of Spain, but more advanced

Table 6.3 Factor content of trade of Greece, Portugal and Spain in trade in manufactures in intra EU-15 trade in 1996

	EU imports from			EU exports to			Net factor content		
	Greece	Portugal	Spain	Greece	Portugal	Spain	Greece	Portugal	Spain
HUMAN CAPITAL-INTENSIVE	19.5	38.3	60.2	57.4	54.8	60.6	−37.9	−16.6	−0.4
High technology	6.8	2.6	7.9	11.9	10.0	12.7	−5.1	−7.4	−4.9
Labour-intensive	4.5	1.9	4.2	4.3	4.4	7.0	+0.2	−2.5	−2.8
Capital-intensive	2.3	0.7	3.7	7.6	5.6	5.8	−5.3	−4.9	−2.1
Medium Technology	9.3	24.3	45.2	38.4	36.5	36.9	−29.1	−12.2	+8.3
Labour-intensive	5.8	6.3	10.3	19.3	17.2	17.4	−13.4	−10.9	−7.1
Capital-intensive	3.5	18.0	34.9	19.2	19.3	19.5	−15.7	−1.3	+15.4
Low technology	3.4	11.4	7.1	7.1	8.3	11.0	−3.7	+3.2	−3.8
LABOUR-INTENSIVE	50.8	50.4	25.4	27.2	30.6	25.6	+23.7	+19.8	−0.2
RESOURCE-INTENSIVE	29.7	11.4	14.4	15.4	14.6	13.8	+14.2	−3.2	+0.6
Human capital:labour ratio	0.38	0.76	2.37	2.11	1.79	2.37	0.18	0.42	1.0
High-tech:labour	0.13	0.05	0.31	0.44	0.33	0.50	0.30	0.16	0.62

Source: estimated from COMEXT database. Methodology derived from Wolfmayr-Schnitzer (1998).

than that of Greece despite the higher per-capita income levels in Greece shown in Chapter 1. The export share of human capital-intensive goods from the CEE-10 to the EU closely resembled that of Portugal and was higher than that of Greece but lower than that of Spain. The net factor content of CEE-10 trade in manufactured goods with the EU in 1996 occupied an intermediate position between Spain's trade with the EU (and intra-EU trade in general) and Greece's trade with the EU. The net factor content of CEE-10 trade with the EU was closest to that of Portugal, although the CEE-10 had a lower surplus in labour-intensive goods which was offset by a surplus in resource-intensive goods which resulted from the higher share of resource-intensive goods in CEE-10 exports to the EU.

6.3.3   A comparison between CEE trade with the EU and selected non-EU countries

Tables 6.4 and 6.5 provide details of the factor content of exports, imports and net trade flows for ten selected emerging market economies in south-east Europe, Latin America and central and south-east Asia with the EU-15. Major differences can be observed between the nature of trade relation of the south-east Asian economies and the EU at one extreme and India and Turkey at the other. Human capital-intensive goods predominate in both south-east Asian exports and imports to and from the EU, ranging from 76.8 per cent of exports from South Korea to the EU to 89.9 per cent of exports from Singapore. South-east Asian exports were also marked by a high proportion of products embodying high-technology (largely in the computing and office equipment categories) which ranged from 22.4 per cent of Japanese exports of manufactured goods to the EU to 55.2 per cent of EU imports from Singapore. Labour-intensive exports from south-east Asia to the EU were also considerably below the world average resulting in relatively high ratios of human capital-intensive exports to labour-intensive exports.

Of the remaining countries included in Table 6.3, only Mexico (56.9 per cent) had a higher proportion of human capital-intensive goods in total exports than the combined CEE-10 (37.9 per cent) with Brazil (35.2 per cent) and China (34 per cent) having comparable levels. Labour-intensive goods accounted for over 60 per cent of the exports of manufactured goods from each of China, Turkey and India compared with 40.8 per cent for the CEE-10, whereas resource-intensive goods were more important than labour-intensive goods in Latin American exports. CEE-10 exports of resource-intensive goods (21.3 per cent) were larger

Table 6.4 Factor content of EU-15 imports of manufactured goods from selected countries in 1996 (percentages of EU imports of manufactured goods from each country)

	Turkey	India	China	Argentina	Brazil	Mexico	South Korea	Malaysia	Singapore	Japan
HUMAN CAPITAL-INTENSIVE	23.6	17.2	34.0	37.9	35.2	56.9	76.8	77.2	89.9	81.2
High technology	7.9	6.5	8.7	14.3	5.6	14.5	27.9	42.8	55.2	22.4
Labour-intensive	7.6	4.5	4.1	5.1	3.7	8.7	19.4	28.3	18.7	14.0
Capital-intensive	0.3	2.0	4.5	9.2	1.9	5.8	8.5	14.5	36.5	8.4
Medium Technology	11.0	9.2	18.9	19.6	19.1	27.4	41.3	24.2	29.9	50.5
Labour-intensive	5.9	3.4	16.0	6.8	7.5	16.8	24.3	21.3	26.5	30.0
Capital-intensive	5.1	5.8	2.9	12.8	11.6	10.6	16.9	2.9	3.4	20.5
Low technology	4.6	1.5	6.4	4.0	10.5	15.0	7.7	10.1	4.8	8.3
LABOUR-INTENSIVE	65.5	69.0	60.6	13.3	24.5	14.6	19.2	17.9	8.4	15.5
RESOURCE-INTENSIVE	10.9	13.8	5.4	48.8	40.3	28.5	3.9	4.9	1.8	3.3
Human capital:labour ration	0.36	0.25	0.56	2.85	1.43	3.90	4.00	4.31	10.72	5.24
High-tech:labour	0.12	0.09	0.14	1.07	0.23	0.99	1.45	2.39	6.58	1.44

Source: estimated from COMEXT database. Methodology derived from Wolfmayr-Schnitzer (1998).

Table 6.5 Factor content of EU-15 exports of manufactured goods and net factor content of selected countries trade with EU in 1996 (percentages of EU exports of manufactures)

	Turkey	India	China	Argentina	Brazil	Mexico	South Korea	Malaysia	Singapore	Japan
EU Exports										
Human capital-intensive	62.0	53.6	75.7	72.3	69.0	62.7	68.3	79.9	75.0	66.1
High technology	13.8	9.5	11.4	11.3	14.8	13.6	14.9	27.9	26.6	16.3
Labour-intensive	24.8	34.2	15.3	19.0	20.6	24.7	18.8	11.0	15.8	25.5
Resource-intensive	13.2	12.2	9.0	8.6	10.4	12.5	12.9	9.1	9.2	8.4
Human capital:labour ratio	2.50	1.57	4.96	3.81	3.35	2.54	3.63	7.28	4.75	2.60
High tech:labour-intensive	0.56	0.28	0.75	0.59	0.72	0.55	0.79	2.55	1.68	0.64
Net factor flows										
Human capital-intensive	−38.4	−36.4	−41.7	−34.4	−33.8	−5.8	+8.5	−2.7	+14.9	+15.1
High technology	−5.9	−3.0	−2.7	−3.0	−9.2	+0.9	+13.0	+14.9	+28.6	+6.1
Labour-intensive	+40.7	+34.8	+45.3	−5.8	+3.9	−10.1	+0.4	+6.9	−7.4	−10.0
Resource-intensive	−2.3	+1.6	+3.6	+40.2	+29.9	+16.0	−8.9	−4.2	−7.4	−5.1
Human capital:labour ratio	0.14	0.16	0.11	0.75	0.43	1.54	1.10	0.59	2.25	2.02
High tech:labour-intensive	0.21	0.32	0.19	1.81	0.32	1.80	1.83	0.94	3.91	2.25

Source: estimated from COMEXT database. Methodology derived from Wolfmayr-Schnitzer (1998). Positive sign indicates positive net exports for partner country.

than those for the remaining countries. The net effect was that the ratio of human capital to labour-intensive exports was far higher from the Latin American economies (ranging from 1.43 for Brazil to 3.9 for Mexico) than for the CEE-10 (0.93) but far lower for India (0.25), Turkey (0.36) and China (0.56). All the economies featured in Table 6.3 had a higher proportion of high-technology products in exports of manufactures to the EU-15 than the combined CEE-10. Despite the assumed CEE demand for machinery and equipment for restructuring and modernisation, only India had a lower proportion of imports of human capital-intensive products from the EU than the combined CEE-10, while all the economies had a higher proportion of imports of goods embodying high-technology. CEE-10 imports of labour-intensive products from the EU (30.6 per cent) were exceeded only by India.

A clear distinction emerges between the net factor content of EU trade flows with the south-east Asian economies on one hand, and the other emerging economies on the other. Japan, Singapore and South Korea, are net exporters of human capital-intensive products while the high level of Malaysian imports of human capital-intensive goods (79.9 per cent) results in a deficit of 2.7 per cent in this category. The net export surplus in human capital-intensive goods from south-east Asia is most marked in the case of high-technology products. However, the net deficit in human capital-intensive goods for Brazil (−33.8 per cent), China (−41.7 per cent), Turkey (−38.4 per cent) and India (−36.4 per cent) was substantially larger than for the CEE-10 (20 per cent). The counterpart of trade in human capital-intensive goods is that the contribution of labour-intensive goods to net surpluses was substantially greater for China (+45.3 per cent), Turkey (+40.7 per cent) and India (+34.8 per cent) than from the CEE-10 (10.2 per cent).

In summary, the technological level of exports of the combined CEE-10 to the EU-15 lagged considerably behind that of the south-east Asian economies and Mexico when measured by the content of human capital-intensive and high technology products. The CEE-10 had similar proportions of human capital-intensive exports to Argentina, Brazil (where resource-intensive exports were significant) and China but lower proportions of high-technology exports than all the countries observed. Higher imports of human capital-intensive goods in China, Argentina and Brazil were the main contributors to higher net inflows than the CEE-10 in this category, while the CEE-10 net outflow of labour-intensive goods (+10.2 per cent) was substantially lower than for China, India or Turkey, but higher than the remaining net exporters of labour-intensive goods (Brazil, Malaysia and South Korea).

6.3   Differences between the factor content of trade flows between individual CEE-10 economies and the EU

The factor content of EU-15 imports to and exports from the individual CEE-10 economies in 1996 are shown in Tables 6.6–6.7 and the net factor content of trade flows between the EU-15 and the individual CEE-10 economies are shown in Table 6.8. Some broad similarities can be observed between the trade patterns of all the CEE-10 economies. No individual economy attained a proportion of exports of human capital-intensive goods, or goods embodying high technology, which attained the level of either intra-EU trade or world exports to the EU. No single economy achieved even half the world average proportion of exports of products embodying high-technology (22.8 per cent) and only Hungary (8.2 per cent) exceeded the proportions of Spain, Turkey and India. The counterpart of this was that each of the individual economies exceeded the proportion of exports of labour-intensive products involved in intra-EU trade (24.7 per cent) and world exports of labour-intensive products to the EU. As a result each of the CEE-10 economies exhibited net factor inflows for human capital-intensive goods. However, these ranged from exceedingly high levels of −40.8 per cent for Latvia and −31.9 per cent for Romania, to modest levels of only – 9.1 per cent for Slovenia and −6.0 per cent for Hungary. Each of the individual CEE-10 economies had net outflows in trade in both labour-intensive and resource-intensive goods. Net outflows of resource-intensive goods were highest for Latvia (+27.7 per cent) and Bulgaria (+26.3 per cent) but were only +0.7 per cent for Hungary. Net outflows of labour-intensive goods were highest for Romania (+22.2 per cent) but were as low as +3.8 per cent for Slovenia.

Significant disparities exist between the net factor contents of the trade flows of the individual CEE-10 economies with the EU which can be largely explained by differences in the structure of CEE-10 exports. The CEE-10 can be divided into three broad groups according to the share of human capital-intensive exports. The most advanced group consists of Hungary (54.4 per cent), Slovenia (44.3 per cent) and the Czech Republic (42.3 per cent) with proportions of human capital-intensive products in exports of manufactured goods which exceed those of Portugal, Argentina, Brazil and China .Each of these countries also has a ratio of human capital-intensive exports to labour-intensive exports of greater than one. Hungary (1.65) had the largest ratio in the region, but this fell behind Spain but was ahead of Brazil.

Table 6.6 The factor content of EU-15 imports of manufactured goods from individual CEE-10 economies in 1996

	Poland	Czech	Hungary	Slovakia	Slovenia	Romania	Bulgaria	Estonia	Latvia	Lithuania
HUMAN CAPITAL-INTENSIVE	32.7	42.3	54.4	34.0	44.3	16.1	22.2	29.9	12.2	21.5
High technology	4.3	5.1	8.2	1.5	4.2	1.2	4.3	2.1	2.4	6.4
Labour-intensive	3.6	4.2	5.7	0.8	3.7	0.6	2.8	1.4	1.8	6.1
Capital-intensive	0.7	0.9	2.5	0.7	0.5	0.5	1.5	0.7	0.6	0.3
Medium technology	22.3	29.9	26.7	24.9	35.0	10.5	15.0	23.9	7.3	6.7
Labour-intensive	10.7	17.7	18.4	9.7	18.2	6.0	7.3	19.3	3.7	3.9
Capital-intensive	11.6	12.2	8.3	15.2	16.8	4.5	7.7	4.6	3.5	2.8
Low technology	6.1	7.3	19.4	7.6	5.1	4.4	2.9	3.8	2.6	8.4
LABOUR-INTENSIVE	43.1	38.5	33.1	36.0	37.0	62.3	38.4	51.2	48.9	46.7
RESOURCE-INTENSIVE	24.2	19.1	12.5	30.0	18.6	21.6	39.4	18.9	38.9	31.9
Human capital:labour ratio	0.76	1.10	1.65	0.94	1.20	0.26	0.58	0.58	0.25	0.46
High tech:labour	0.10	0.13	0.25	0.04	0.11	0.02	0.11	0.04	0.05	0.14
TOTAL (ECU mn)	9 810	8 303	7 136	3 065	4 068	3 326	1 344	673	329	683

Source: estimated from COMEXT database. Methodology derived from Wolfmayr-Schnitzer (1998).

Table 6.7 The factor content of EU-15 exports of manufactured goods to individual CEE-10 economies in 1996

	Poland	Czech	Hungary	Slovakia	Slovenia	Romania	Bulgaria	Estonia	Latvia	Lithuania
HUMAN CAPITAL-INTENSIVE	57.4	62.3	60.4	60.0	53.4	48.0	52.6	54.6	53.0	55.1
High technology	8.9	10.4	10.6	9.2	6.9	7.2	8.6	8.1	7.9	9.8
Labour-intensive	3.7	5.7	6.0	4.4	4.2	3.2	3.7	4.5	2.8	5.4
Capital-intensive	5.2	4.7	4.6	4.8	2.7	4.0	4.9	3.5	5.0	4.4
Medium technology	41.5	43.9	38.0	40.8	38.0	34.1	38.4	40.0	36.6	38.6
Labour-intensive	23.0	26.7	25.1	24.2	17.2	22.7	20.3	24.8	21.1	21.0
Capital-intensive	18.5	17.2	12.9	16.6	20.8	11.4	17.1	15.2	15.5	17.6
Low technology	6.9	7.9	11.8	10.0	8.6	6.7	5.5	6.6	8.5	6.7
LABOUR-INTENSIVE	30.8	27.1	27.8	30.0	33.2	40.1	34.2	34.5	35.8	33.3
RESOURCE-INTENSIVE	11.8	10.6	11.8	9.9	13.3	11.9	13.2	10.8	11.2	11.6
Human capital:labour ratio	1.86	2.30	2.18	2.00	1.61	1.20	1.54	1.58	1.48	1.65
High tech:labour	0.29	0.39	0.38	0.31	0.21	0.18	0.25	0.23	0.22	0.29
TOTAL (ECU mn)	16 959	12 378	9 144	3 801	1 435	4 652	3 532	1 305	734	1 161

Source: estimated from COMEXT database. Methodology derived from Wolfmayr-Schnitzer (1998).

Table 6.8 Net factor content of CEE-10 trade with EU-15 in manufactured goods in 1996

	Poland	Czech	Hungary	Slovakia	Slovenia	Romania	Bulgaria	Estonia	Latvia	Lithuania
HUMAN CAPITAL-INTENSIVE	−24.7	−20.0	−6.0	−26.1	−9.1	−31.9	−30.4	−24.8	−40.8	−33.7
High technology	−4.7	−5.3	−2.4	−7.7	−2.7	−6.0	−4.3	−5.9	−5.5	−3.4
Labour-intensive	−0.1	−1.5	−0.3	−3.6	−0.5	−2.5	−0.9	−3.1	−1.0	+1.6
Capital-intensive	−4.6	−3.8	−2.1	−4.1	−2.2	−3.5	−3.3	−2.8	−4.5	−5.0
Medium Technology	−19.2	−14.0	−11.3	−16.0	−3.0	−23.6	−23.5	−16.1	−29.4	−31.9
Labour-intensive	−12.3	−8.9	−6.7	−14.5	+1.0	−16.7	−13.0	−5.4	−17.3	−17.1
Capital-intensive	−6.9	−5.1	−4.6	−1.5	−4.0	−6.9	−10.4	−10.7	−12.0	−14.8
Low technology	−0.8	−0.6	+7.7	−2.4	−3.4	−2.3	−2.6	−2.8	−5.9	+1.7
LABOUR-INTENSIVE	+12.3	+11.4	+5.3	+6.0	+3.8	+22.2	+4.1	+16.7	+13.1	+13.4
RESOURCE-INTENSIVE	+12.4	+8.5	+0.7	+20.1	+5.3	+9.7	+26.3	+8.1	+27.7	+20.3
Human capital:labour ratio	0.41	0.48	0.76	0.47	0.75	0.22	0.38	0.37	0.17	0.28
High tech:labour	0.34	0.34	0.65	0.14	0.55	0.10	0.45	0.18	0.22	0.47

Source: estimated from COMEXT database. Methodology derived from Wolfmayr-Schnitzer (1998).

The share of high-technology sectors in exports of manufactures was unexceptional and even that of Slovenia (4.2 per cent) and the Czech Republic (5.1 per cent) was lower than that of a range of countries which included China, Brazil, India, Turkey and Greece. Poland, Slovakia and Estonia fall into an intermediate category with human capital-intensive products accounting for 32.7 per cent, 34.0 per cent and 29.9 per cent of exports of manufactured goods respectively, a proportion below that of Portugal, but substantially higher than Greece and similar to that of China and Brazil. However, the export structures of Estonia (2.1 per cent) and Slovakia (1.5 per cent) also reflect a very low proportion of products embodying high-technology that place them at less than one-tenth of the world average for exports to the EU in this category resulting in a ratio of high-technology intensive exports to labour-intensive exports of 0.04 which was less than half the level for India. Estonia (51.2 per cent) and Poland (43.1 per cent) also had a high proportion of labour-intensive exports. Consequently the ratio of human capital-intensive exports to labour-intensive exports fell to 0.76 for Poland (comparable to Portugal) and 0.58 for Estonia (higher than Greece and comparable to China).

The final group consists of economies with shares of human capital-intensive exports that are substantially less than half the intra-EU level of 60 per cent. Bulgaria (22.2 per cent), Lithuania (21.5 per cent) had levels of human capital-intensive exports that exceed that of Greece (19.5 per cent) and India (17.2 per cent), but Romania (16.1 per cent) and Latvia (12.2 per cent) fell below even these levels. Romania (62.3 per cent) and the Baltic states are the most dependent on labour-intensive exports resulting in ratios of human capital-intensive exports to labour-intensive exports of 0.26 for Romania and 0.25 for Latvia. Bulgaria had a high level of resource-intensive exports and a high level of high technology exports (in comparison with other CEE-10 economies) which resulted in ratios of human capital-intensive and high-technology exports to labour-intensive exports to 0.58 and 0.11 respectively.

The effect on the factor content of net exports is that the Black Sea economies (Romania and Bulgaria) and the Baltic States (Estonia, Latvia and Lithuania) are all marked by high levels of net inflows of goods embodying human capital and, with the exception of Bulgaria, by high outflows of labour-intensive products. Three central European states (Poland the Czech Republic and Slovakia) are marked by intermediate net inflows of human capital-intensive products which are balanced by a combination of labour and resource-intensive outflows. Only Hungary and Slovenia fall into a category where net inflows of human capital products fall under 10 per cent.

6.4 Changes in the factor content of CEE exports to the EU in the transition to a market economy

A comparison of the factor content of EU imports of manufactured goods from the five CMEA economies in 1988 and 1996 in Tables 4.10 and 6.5 indicates that the proportion of human capital-intensive goods in EU imports has increased in the case of all the economies concerned (treating the Czech and Slovak economies as a single entity) with the exception of Bulgaria where the proportion of human capital-intensive exports has fallen from 32.6 per cent to 22.2 per cent. Hungary has experienced a dramatic improvement in the factor content of its exports with the proportion of human capital-intensive goods rising from 29.3 per cent in 1988 to 54.4 per cent in 1996, while goods embodying high technology have grown from 5 per cent to 8.2 per cent. Poland has experienced a very modest increase with the proportion rising from 31.9 per cent to only 32.7 per cent. These comparisons suffer from two major disadvantages. Firstly, the composition of the EU changed in 1995 when Austria, Finland and Sweden became members, which has not only affected the structure of EU imports from the CEEs, but has affected the structure of imports from individual countries differently. Secondly, EU imports from Poland have grown far more quickly than imports from Hungary, reflecting Poland's relative success in increasing exports of labour-intensive goods. This reflects the dual nature of the Polish economy and has had the effect of biasing the *proportion* of exports of human capital-intensive goods downwards. Changes in the factor content of EU imports from the CMEA-5 between 1988 and 1994 will be examined to overcome the problem of EU enlargement in 1995.[2] The analysis of changes in the structure of trade will be supplemented by an analysis of the differences in the growth in value of exports in each sector for the individual countries.

The factor content of the structure of EU imports from the former CMEA-5 economies are shown in Table 6.9. Data on EU imports from the Czech Republic and Slovakia have been aggregated to form a single country for comparison with Czechoslovakia in 1988. Some differences between the data for 1994 and 1996 are apparent. In general, the improvement in the factor content of EU imports from the CMEA-5 between 1988 and 1994 is not as good as that between 1998 and 1996. This partly reflects changes in the structure of EU imports from the outside world following the entry to the EU of Sweden, Austria and Finland , and partly the continuing improvement in the commodity structure of EU imports from the CEE economies between 1994 and 1996. Changes in the percentage shares of each factor-content sector between 1988 and 1994 are shown in Table 6.10, where a negative sign indicates that the sector concerned fell as a share of EU imports between 1998 and 1994. The share of human capital-intensive goods from the CMEA-5 increased by 4.06 per cent from 26.12 per cent in 1988 to 30.18 per cent in 1994 with goods embodying high technology growing by 1.18 per cent to 4.01 per cent. The share of labour-intensive goods also rose by 2.39 per cent, while the share of resource-intensive goods fell by 6.45 per cent. However the overall increase in the proportion of human capital-intensive goods resulted entirely from the growth of EU imports from Hungary and the Czech and Slovak republics while the share of human capital goods in imports from Bulgaria, Romania and Poland fell. All countries except Hungary, experienced a rise in the share of labour-intensive exports and a fall in the share of resource-intensive exports. Table 6.11 shows the increase in the value of EU imports under each sector between 1988 and 1994 and Table 6.12 shows the share of each sector in the growth of EU imports over the period. The signs in all the cells for Tables 6.11 and 6.12 are all positive. Table 6.12 shows that EU imports of labour-intensive goods accounted for 44.29 per cent of the

Table 6.9   The technological level of exports of manufactured goods from former CMEA economies to the EU-12 in 1994

	Bulgaria	Czech & Slovakia	Hungary	Poland	Romania	Total
Human capital-						
intensive	25.15	34.16	41.70	26.44	11.94	30.18
High technology	8.34	3.84	5.77	3.37	1.60	4.01
labour-intensive	6.16	2.88	4.33	2.92	1.05	3.12
capital-intensive	2.18	0.97	1.43	0.45	0.55	0.89
Medium technology	12.46	24.59	21.57	18.31	6.71	19.49
labour-intensive	6.02	13.40	13.94	8.16	4.00	10.42
capital-intensive	6.43	11.19	7.63	10.15	2.71	9.07
Low technology	4.36	5.73	14.36	4.76	3.64	6.69
Labour-intensive	36.11	38.16	42.06	43.64	63.05	43.22
Resource-intensive	38.74	27.68	16.25	29.93	25.01	26.60
Total (mn ECU)	1 015	7 039	3 813	6 937	2 239	21 045

Source: estimated from COMEXT database. Methodology derived from Wolfmayr-Schnitzer (1988).

Table 6.10   Changes in the share of the factor content of EU-12 imports of manufactured goods from former CMEA between 1988 and 1994 (per cent)

	Bulgaria	Czech & Slovakia	Hungary	Poland	Romania	Total
Human capital-						
intensive	−7.41	+6.81	+12.36	−5.42	−1.00	+4.06
High technology	+1.83	+1.44	+0.82	+1.1	+0.18	+1.18
Medium technology	−8.96	+3.54	+0.18	−1.04	−3.00	+1.35
Low technology	−0.26	+1.82	+11.36	−5.49	+1.84	+1.53
Labour-intensive	+8.33	+4.48	−2.25	+9.09	+6.87	+2.39
Resource-intensive	−0.93	−11.29	−10.11	−3.66	−5.87	−6.45

Source: estimated from COMEXT database. Methodology derived from Wolfmayr-Schnitzer (1998).

Table 6.11   Changes in the value of exports of manufactures from former CMEA economies to the EU-12 by factor content between 1988 and 1994 (ECU mn)

	Bulgaria	Czech & Slovakia	Hungary	Poland	Romania	Total
Total (ECU mn)	754	5 477	2 485	5 023	810	14 551
% growth	287	351	187	263	57	224
Human capital-						
intensive	170	1 977	1 200	1 224	82	4 655
High technology	68	233	154	190	15	660
Medium technology	70	1 402	539	900	11	2 923
Low technology	32	342	508	134	56	1 072
Labour-intensive	294	2 160	1 015	2 366	609	6 445
Resource-intensive	290	1 340	270	1 433	119	3 451

Source: estimated from COMEXT database. Methodology derived from Wolfmayr-Schnitzer (1998).

increase in the value of imports over the period and imports of capital-intensive goods for 31.99 per cent. Although the share of EU imports of human capital-intensive goods from Hungary rose more rapidly than from other former CMEA economies, this was partly attributable to the slow rate of growth of imports of resource-intensive goods, which was reflected in the slower overall rate of growth of EU imports from Hungary. EU imports of human capital-intensive goods from the Czech and Slovak republics per head of the exporting populations actually grew faster (ECU 126 per capita) than from Hungary (ECU 118 per capita).

Furthermore the growth in value of EU imports of human capital-intensive goods from the Czech and Slovak republics was greater in the sectors embodying high and medium technology. Unfortunately, it is not possible to separate the *growth* in EU imports between the Czech and Slovak republics as comparable data are not available for 1988, although data on the structure of trade in 1994 show that EU imports from Czech Republic had a higher content of human capital-intensive goods and high technology goods than imports from Slovakia.

Poland's export performance is slightly more impressive when examined from the perspective of absolute changes in value of EU imports as opposed to shares. The fall in the *share* of EU imports of human capital-intensive goods can be partly attributed to a higher growth of EU imports of labour-intensive goods from Poland, which reflects the successful diversion of Polish exports of textiles and clothing from the Soviet to the EU market, and partly to the slower rate of growth of imports of human capital-intensive goods.

The gap between the export performance of the central east European states and the Balkan states, and Romania in particular, is marked. EU imports of manufactured goods from Romania grew by only 57 per cent between 1988 and 1994[3] while the increase in the value of EU imports of human capital-intensive good from Romania was only ECU 82 million, equivalent to less than four ECU per head. Labour-intensive goods, the majority of which were produced under outward processing arrangements (clothing, furniture and footwear) accounted for 75.17 per cent of the growth of EU imports from Romania. Although EU imports from Bulgaria grew by 287 per cent between 1998 and 1994, this was from a

Table 6.12   Source of growth of exports of manufactured goods from former CMEA economies to the EU-12 between 1988 and 1994 (per cent)

	Bulgaria	Czech & Slovakia	Hungary	Poland	Romania	Total
Human capital-intensive	22.58	36.10	48.30	24.37	10.18	31.99
High technology	8.98	4.25	6.20	3.78	1.91	4.54
Medium technology	9.34	25.60	21.67	17.92	1.40	20.09
Other	4.26	6.24	20.43	2.67	6.87	7.36
Labour-intensive	39.00	39.44	40.85	47.10	75.17	44.29
Resource-intensive	38.42	24.46	10.85	28.53	14.64	23.72

Source: estimated from COMEXT database. Methodology derived from Wolfmayr-Schnitzer (1998).

very low starting point. The growth in imports was more heavily con centrated on labour-intensive and resource-intensive goods (iron and steel and non-ferrous metals). The growth of EU imports of chemicals from Bulgaria was small and was concentrated on low value fertilisers and inorganic chemicals which was disappointing in view of this sector's importance in Bulgaria's export structure in the communist period.

6.5 Conclusions

The evidence derived from tests of the factor content of trade between the EU and the CEE-10 in 1996 indicates that a substantial economic distance exists between the CEE-10 economies as a whole and the EU-15. Major differences exist between the technological level and factor con- tent of trade between the combined CEE-10 and the EU-15 and the general pattern of intra-EU trade. Differences in the net factor content of trade between the CEE-10 and the EU and intra-EU trade, and trade between the rest of the world and the EU can be largely attributed to differences in the structure of CEE-10 exports rather than imports. CEE- 10 exports to the EU-15 contained a significantly lower proportion of exports of human capital-intensive goods than was the norm in either intra-EU trade or EU imports from the rest of the world and correspond- ingly higher proportions of exports of labour-intensive and resource- intensive products. CEE-10 exports to the EU were also marked by a significantly lower proportion of commodities embodying high-tech- nology than intra-EU trade or EU imports from the rest of the world. The technological level of combined CEE-10 exports to the EU was substantially less advanced than that of Spain, on a roughly similar level to that Portugal , but was more advanced than that of Greece. The technological structure of exports was substantially less advanced than that of the newly industrialising economies of south-east Asia, but on a similar level to Brazil, Argentina and China but more advanced than India or Turkey. Although the technological content of exports of none of the CEE economies attained the levels of either intra-EU trade or EU imports from the rest of the world, substantial differences existed between them. Hungary, Slovenia and the Czech Republic were the most advanced, followed by an intermediate group consisting of Slovakia, Poland and Estonia, while Bulgaria, Lithuania, Romania and Latvia fall into a final category with low proportions of human capital-intensive exports.

7
Revealed Specialisation and the Competitiveness of the CEE Economies

7.1 Introduction

It was established in Chapter 6 that the exports of the CEE-10 to the EU-15 contained significantly higher proportions of labour-intensive and resource-intensive goods and lower proportions of human capital-intensive goods than were contained in the general structure of both intra-EU trade and EU imports from the rest of the world. However, this indicates that the CEE economies were capable of penetrating EU markets in some human capital-intensive sectors, including sectors embodying high-technology, which could become the basis of export expansion in the future. Resource intensive and labour-intensive sectors also embrace a wide range of different industries, with different characteristics and implications for existing EU producers and economic restructuring in the CEE. Which sectors of industry are the most likely to expand in the CEE economies and which to contract if industrial concentration takes place in Europe as predicted in Chapter 2? Will the liberalisation of trade relations between the EU and CEE-10 result in increased imports from CEE-10 economies at the expense of imports from the rest of the world or will the removal of trade barriers result in the displacement of existing EU producers, and if so, in which industrial sectors will this be most important? In which specific industrial sectors will trade liberalisation leave the domestic industries of the CEE economies vulnerable to competition as they open their markets to competition from EU manufacturers?

This Chapter will provide a framework for answering these questions by establishing the specific industrial sectors in which the CEE economies have demonstrated that they are capable of withstanding competitive pressures in trade in the EU market in competition with both existing

EU producers and with other exporters to the EU. It will also indicate the industrial sectors in which the CEE economies may be vulnerable to increased competition from producers in the EU. The analysis will be based on indices of revealed comparative advantage (RCA) and export specialisation (ESI) which were described in Chapter 3. Tests have been conducted at the two and three-digit levels of the Standard International Trade Classification (SITC) for trade in manufactured goods (SITC categories 5–8) between the combined CEE-10 and each individual CEE economy and the EU-15 for 1996. RCA indices have been estimated for bilateral trade between the EU-15 and the CEE-10. ESI indices have been estimated by comparing the structure of EU-15 imports from the CEE-10 with EU-15 imports from the world, excluding intra-EU trade.

One of the concerns raised about the use of RCA and ESI indices to predict future trade flows is that they can only measure past trade flows which have been affected by non-market criteria which will not apply after full trade liberalisation (for example quotas, voluntary restraints, tariffs and subsidies). One of the major problems relating to trade between the EU and the former centrally-planned economies is that investment decisions were taken by central planners for essentially non-market reasons which has resulted in the development of export sectors which were not based on considerations of comparative advantage. It is also possible that planners failed to develop industrial sectors in which the CEE economies may be capable of competing in the long-term for ideological reasons (for example, industries supplying the publishing industry) or failure to recognise the market potential (for example, passenger cars). It would be expected that export patterns that are not based on a cost advantages reflecting comparative advantage would have been progressively reduced as a result of price liberalisation, the removal of hidden and open subsidies and the reduction and removal of EU quotas and tariffs and the exposure of CEE industries to international competition and would be of reduced importance by the mid 1990s. Another possibility is that investment decisions that were initially based on non-economic factors may have generated internal or external economies (for example, a specialised labour force, training facilities, transport facilities, sources of hydroelectric power) that have created a genuine and enduring comparative advantage that did not previously exist. On the other hand, a high level of imports in an undeveloped sector could still be an indication of a short-term inability to satisfy domestic demand from domestic production which could be the precursor of the growth of domestic output with the potential for exports. It was also argued in Chapter 3, that an industrial sector needed to satisfy the criteria established by both the RCA and the ESI tests before it could be considered to be capable of withstanding competition from both existing manufacturers inside the EU and from competitors outside the EU. The term *revealed specialisation* will be used to describe those industrial sectors in which the combined CEE-10 or an individual CEE-10 economy has a positive RCA index and an ESI greater than one. Industrial sectors that satisfy both these criteria offer the best prospects for trade creation as trade between the two regions is liberalised.[1]

One of the most important considerations from the perspective of EU policy makers will be the impact enlargement will have on the aggregate supply and demand for specific sectors in the EU as a whole. The potential for increased EU imports of agricultural machinery from a single country (for example Hungary, which has a revealed specialisation in this sector) may be far less important for EU producers than the potential for the growth of EU exports of agricultural machinery to other CEE markets. The results of the RCA and ESI tests will first be presented at the level of trade between the combined CEE-10 and the EU to provide an indication of total trade effects. ESI and RCA indices by themselves do not indicate the size of trade flows and high indices can be recorded on small volumes of trade. The size of EU-15 exports to and imports from the combined CEE-10 and the resulting trade balances are shown for each industrial sector together with the RCA and ESI indices. This also provides a visual indication of those sectors in which export trade is important for the CEE-10, but in which they face severe competition from other products in the EU or from competition outside the EU. Although trade data are presented from the perspective of the EU, the RCA and ESI indices have been estimated, and presented, from the perspective of the CEE economies. Finally, the sectors in which individual CEE economies demonstrated a revealed specialisation are presented in separate tables to help identify sectors in which individual economies may be capable of expanding exports. Countries have been ranked according to the size of their ESI in these tables. This provides an indication of their export competitiveness in EU markets, in relation to one another. The analysis has been broken down into five basic industrial sectors with subdivisions. These consist of chemical products (SITC 5) which has been broken down into organic and inorganic chemicals and fertilisers (SITC 51, 52 and 56) and other chemicals (SITC 53–5, 57–9); machinery and equipment (SITC 7) which has been subdivided into general industrial machinery (SITC 70–4), office and telecommunications equipment (75–7) and transport equipment (78–9); manufactured goods classified by material (SITC 6) which has been subdivided into manufactures of leather, wood, paper and textiles (SITC 60–5) and minerals and metals (SITC 66–9); light industrial consumer goods (SITC 82–5) and other miscellaneous manufactures (SITC 81 and 87–9).

7.2 Comparative advantage and revealed specialisation in trade in chemical products

7.2.1 Patterns of trade in chemicals between the EU and the combined CEE-10

The EU made a substantial surplus in trade in chemical products with the CEE-10 of ECU 4560 million out of a surplus of ECU 43,406 million in chemicals in total extra-EU trade in 1996. This resulted in an RCA index for the CEE-10 of −0.277 for trade with the EU in chemicals as a whole (see Table 7.1). The low level of competitiveness of the CEE-10 economies in relation to exporters from outside the EU, which are dominated by the USA, Japan and Switzerland, was shown by the ESI of 0.700. Trade in chemical products between the EU-15 and the CEE-10 can be divided into two distinct categories. The EU was a substantial net importer of organic chemicals (SITC 51), inorganic chemicals (SITC 52) and fertilisers (SITC 56) and a net exporter of all other chemical products

Table 7.1 EU-15 trade with the CEE-10 in organic and inorganic chemicals and fertilisers in 1996 (ECU million)

	EU export	EU import	Balance	CEE RCA	CEE ESI
Total chemicals (5)	7 579	3 018	4 560	−0.277	0.700
Organic chemicals (51)	678	738	−60	+0.215	0.633
Hydrocarbons & derivatives (511)	116	128	−12	+0.221	1.063
Alcohols, phenols & derivs (512)	92	109	−17	+0.254	0.853
Carboxylic acids (513)	98	119	−21	+0.269	0.846
Nitrogen-function compounds (514)	142	70	−27	+0.258	0.744
Organo-inorganic compounds (515)	125	158	−33	+0.285	0.349
Other organic chemicals (516)	105	55	50	−0.149	0.554
Inorganic chemicals (52)	249	424	−175	+0.415	1.018
Inorganic chemical elements (522)	108	257	−148	+0.542	1.321
Metal salts and peroxysalts (523)	106	129	−23	+0.268	1.809
Other inorganic chemicals (524)	13	35	−22	+0.588	0.882
Radioactive and assoc (525)	22	4	18	−0.615	0.034
Fertilisers (56 and 562)	71	631	−560	+0.853	3.418

Source: estimated from COMEXT database.

at the SITC two-digit level. The combined CEE-10 recorded a revealed specialisation in only seven of the 33 SITC three-digit categories in chemicals. Trade in organic chemicals (SITC 51) is a key sector in which the CEE-10 recorded a positive RCA index (+0.215) and an ESI of less than one (0.633) indicating the possibility of trade diversion. The EU was a major net importer of inorganic chemicals from outside the EU but CEE-10 exporters accounted for only 6.1 per cent of EU imports in this sector.

Table 7.2   EU-15 trade with the CEE-10 in chemical products other than organic and inorganic chemicals in 1996 (ECU million)

	EU export	EU import	Balance	CEE RCA	CEE ESI
Dyeing, tanning and colouring materials (53)	898	92	806	−0.745	0.413
Synthetic organic colouring (531)	109	43	66	−0.284	0.374
Dyeing and tanning extracts (532)	20	3	18	−0.668	0.329
Pigments, paints, varnishes (533)	768	47	722	−0.841	0.466
Medicinal and pharmacy (54)	1 418	140	1 278	−0.754	0.151
Other medicinal products (541)	319	103	216	−0.372	0.208
Medicaments (542)	1 099	37	1 062	−0.909	0.086
Essential oils, resinoids, perfumes (55)	916	60	856	−0.831	0.283
Essential oils and materials (551)	127	11	116	−0.780	0.161
Perfumes, cosmetics (553)	524	28	496	−0.860	0.273
Soaps, cleansing, polishes (554)	265	21	244	−0.798	0.515
Plastics in primary form (57)	1 062	588	473	−0.118	**1.254**
Polymers of ethylene (571)	112	178	−66	+0.387	**2.038**
Polymers of styrene (572)	110	68	42	−0.063	**1.971**
Polymers of vinyl chloride (573)	115	158	−43	+0.323	**3.234**
Polyacetals (574)	247	42	205	−0.611	0.390
Other plastics in primary form (575)	473	136	337	−0.418	0.730
Waste and scraps (579)	3	4	−1	+0.331	**1.418**
Plastics in non-primary form (58)	1 080	175	905	−0.626	0.727
Tubes, pipes, hoses (581)	263	65	198	−0.482	**1.561**
Plastic plates, sheets, film, foil (582)	686	105	581	−0.642	0.540
Monofilaments (583)	131	5	126	−0.897	**1.264**
Other chemical products (59)	1 207	169	1 038	−0.668	0.357
Insecticides, herbicides (591)	325	32	293	−0.758	0.463
Starches, albuminoidal (592)	140	61	80	−0.239	0.855
Explosives, pyrotechnics (593)	5	15	−10	+0.622	0.830
Additives for mineral oils (597)	165	4	161	−0.933	0.115
Miscellaneous chemicals (598)	572	57	515	−0.751	0.204

Source: estimated from COMEXT database.

EU imports of organic chemicals from outside the EU amounted to ECU 12,033 million in 1996 and were dominated by imports from OECD countries including the USA (ECU 3752 million), Switzerland (ECU 2538 million) and Japan (ECU 1273 million). The Czech Republic, which was the leading exporter from the CEE-10 with exports of ECU 206 million, was only the ninth-largest exporter to the EU. The combined CEE-10 recorded a revealed specialisation in SITC 511 (hydrocarbon derivatives) and positive RCA indices combined with ESI indices of less than one in four of the remaining five three-digit categories (SITC 512–15). The average unit value of CEE exports of organic chemicals was 52.6 per cent of the average unit value of EU imports indicating that the CEE economies were largely competing in the lower quality ranges of the market. The combined CEE-10 recorded a comparative disadvantage measured by both RCA and ESI indicators in 17 of the 33 three-digit sectors in chemical production. These all came under the human capital-intensive categories, the majority of which embodied high technology.

Table 7.3 provides an indication of the relative quality levels of EU exports to the CEE-10 and EU imports from the CEE-10 in trade in chemical products in 1996. Table 7.3 shows that while 93.1 per cent of EU exports of chemicals to the CEE-10 consisted of products which could be classified as human capital-intensive according to the Wolf-mayr-Schnitzer categories, only 61.7 per cent of EU imports from the CEE-10 were classified as human capital-intensive, while the remainder were classified as resource-intensive. Furthermore, only 15.6 per cent of CEE-10 exports to the EU were classified as high technology compared with 40 per cent of EU exports to the CEE-10. The net effect was that the EU made a surplus in trade in human capital-intensive chemical

Table 7.3 Factor content of trade between EU-15 and CEE-10 in chemicals in 1996 (ECU million)

	EU exports		EU imports		Balance
	ECU mn	%	ECU mn	%	ECU mn
Total	7 576	100.0	3 018	100.0	+4 558
Human capital-intensive	7 057	93.1	1 861	61.7	+5 196
high technology	3 029	40.0	472	15.6	+2 557
medium technology	3 094	40.8	1 264	41.9	+1 830
low	934	12.3	125	4.2	+809
Resource-intensive	519	6.9	1 157	38.3	−638

Source: estimated from COMEXT database.

products of ECU 5196 million and a deficit in trade in resource-intensive chemical products of ECU 638 million.

7.2.2 Revealed specialisation in trade in chemicals by individual CEE economies

Table 7.4 indicates that the Czech republic had a revealed specialisation in the highest number of sectors (8), followed by Hungary (7), Slovakia (6) and Bulgaria (5). The revealed comparative advantage for the combined CEE-10 in hydrocarbon derivatives (SITC 511) was entirely attributed to exports from the Czech Republic and Hungary. All CEE-10 economies, except Slovakia and Romania, had a revealed specialisation in SITC 522 (inorganic chemical elements) while Bulgaria was the dominant CEE-10 exporter of metal salts and peroxysalts (SITC 523). All the CEE-10 except Hungary and Slovenia recorded a revealed specialisation in trade in fertilisers (SITC 562) where the CEE-10 accounted for just

Table 7.4   CEE-10 economies with revealed specialisation in chemical products

SITC number and description	Countries recording positive RCA index and ESI > 1, ranked by ESI
Hydrocarbons & derivatives (511)	Cz (2.3) Hun (2.2)
Alcohols, phenols & derivatives (512)	Bul (4.0) Lat (3.4) Sla (1.6) Pol (1.3)
Carboxylic acids (513)	Cz (1.8) Sle (1.2)
Nitrogen-function compounds (514)	Lat (1.6) Hun (1.4) Sla (1.2)
Other organic chemicals (516)	Sla (1.1) Bul (1.1)
Inorganic chemical elements (522)	Est (6.8) Lat (4.4) Lit (3.5) Bul (2.1) Hun (1.6) Sle (1.5) Pol (1.3) Cz (1.0)
Metal salts and peroxysalts (523)	Bul (17.4) Pol (2.6) Sle (1.4) Rom (1.0)
Other inorganic chemicals (524)	Rom (2.2) Sle (2.2) Sla (1.3)
Radioactive and associated (525)	Est (1.3)
Synthetic organic colouring (531)	Cz (1.4)
Chemical fertilisers (562)	Lit (39.5) Bul (15.1) Est (6.4) Rom (4.3) Sla (4.2) Lat (4.1) Pol (3.6) Cz (1.8)
Polymers of ethylene (571)	Sla (7.2) Bul (5.3) Hun (3.8) Cz (2.0)
Polymers of styrene (572)	Hun (5.3) Cz (4.2)
Polymers of vinyl chloride (573)	Rom (8.1) Hun (4.2) Cz (3.7) Sla (3.6) Pol (2.7)
Insecticides, herbicides (591)	Sle (1.1)
Starches, albuminoidal (592)	Lit (14.9) Lat (7.4)
Explosives, pyrotechnics (593)	Cz (1.93)

Abbreviations: Bulgaria (Bul), Czech Republic (Cz), Hungary (Hun), Poland (Pol), Romania (Rom), Slovakia (Sla), Slovenia (Sle), Estonia (Est), Latvia (Lat) Lithuania (Lit).
Source: estimated from COMEXT database.

under a third of EU imports of fertilisers of ECU 1919 million in 1996. The Baltic states and Slovenia generally displayed a low export capability in plastics in primary form. No country had a comparative advantage in any of the three-digit sectors comprising categories SITC 54 (medicinal products and medicaments) and SITC 55 (essential oils and perfumes) and SITC 58 (plastics in non-primary forms) which generally command higher unit values in CEE markets. The failure of exporters from Bulgaria, Hungary and Poland to penetrate EU markets in medicines and cosmetics is disappointing in view of the importance of exports from these sectors to the Soviet Union in 1988.

7.3 Comparative advantage and revealed specialisation in trade in machinery and equipment

7.3.1 Patterns of trade in machinery and equipment between the EU and the CEE-10

Trade in machinery and equipment accounted for 41 per cent of EU-15 exports to the CEE-10 and for 28 per cent of EU-15 imports from the CEE-10 in 1996. The EU made a surplus of ECU 12,950 million in trade in total trade machinery and equipment (SITC 7) with the CEE-10 resulting in an RCA index for the CEE-10 of −0.155 (see Table 7.6). This reflects the continuation of CEE deficits in trade in machinery and equipment with the EU from the communist era and the continued demand for imported equipment for industrial modernisation and restructuring. The ESI of 0.731 indicates that the CEE-10 have been less successful in penetrating EU markets for machinery and equipment than other extra-EU exporters, despite the fact that the CMEA economies recorded major surpluses in trade in machinery and equipment with the Soviet Union in the communist era. The EU-15 recorded visible trade surpluses in all SITC two-digit categories in machinery and equipment with the CEE-10 except SITC 71 (power generating equipment) and SITC 79 (other transport equipment). The combined CEE-10 recorded a revealed specialisation in only 12 out of the 51 three-digit SITC categories in machinery and equipment and a comparative disadvantage by both indices in 19 categories.

Trade in machinery and equipment between the EU and the CEE-10 in both directions has a substantially smaller component of trade in commodities involving high technology than the structure of EU trade with other non-EU partners and intra-EU trade. Table 7.5 shows that, although the values of EU exports of machinery and equipment to the

Table 7.5   Factor content of EU trade with CEE-10 in machinery and equipment

	EU trade in machinery with CEE-10 (ECU mn)			Percentage breakdown of EU trade in machinery		
	EU exports	EU imports	Balance	Exports to CEE	Imports from CEE	Imports from world
Human capital-intensive	21 771	11 088	10 683	83.3	84.1	91.6
High technology	2 406	1 378	1 028	9.2	10.5	37.1
labour-intensive	1 720	1 194	526	6.6	9.1	26.2
capital-intensive	686	184	502	2.6	1.4	11.0
Medium technology	16 065	6 592	9 473	61.5	50.0	43.6
labour-intensive	11 732	4 260	7 471	44.9	32.3	35.9
capital-intensive	4 333	2 332	2 002	16.6	17.7	7.7
Low technology	3 300	3 118	182	12.6	23.6	10.9
Labour-intensive	4 365	2 098	2 267	16.7	15.9	8.4
Total	26 137	13 187	12 950	100.0	100.0	100.0

Source: estimated from COMEXT database.

CEE-10 is nearly double the value of imports of machinery from the CEE-10, the factor content of EU imports not very dissimilar from the factor content of EU exports. Indeed the *proportion* of EU imports of machinery from the CEE-10 which embody high technology (10.5 per cent) is higher than the proportion of these goods in EU exports to the CEE-10 (9.2 per cent) but substantially less than in intra-EU trade (21.1 per cent) indicating that this reflects a low absorptive capacity by the CEE-10 economies for machinery and equipment embodying high-technology. However, a major difference can be observed between the structure of EU imports of machinery from the world as a whole and from the CEE-10, where products which embody high-technology account for 37.1 per cent of machinery imports, with compensating lower proportions of medium and low-technology products. This is explained by EU imports of office equipment and computers, valves, transistors and aircraft from the USA and south-east Asia.

Trade in machinery and equipment encompasses highly differentiated technologies and processes. Consequently, the analysis for the remainder of this section will be divided into three separate sections; trade in machinery and equipment for general industrial purposes; trade in equipment for offices and telecommunications equipment and electrical equipment, and trade in transport equipment.

7.3.2 Trade in machinery and equipment for general industrial purposes (SITC categories 70–4)

Table 7.6 shows that the combined CEE-10 recorded a revealed specialisation in six three-digit industrial sectors in the SITC categories 70–4.

Table 7.6 EU-15 trade with the CEE-10 in machinery and equipment for general industrial purposes in 1996 (ECU million)

	EU export	EU import	Balance	RCA	ESI
Total machinery and equipment (7)	26 137	13 186	12 950	−0.155	0.731
Total (70–4)	10 393	4 382	6 011		
Complete industrial plant (70, 700)	145	0	145	−1.000	0.000
Power generating (71)	1 384	1 641	−257	+0.256	1.105
Steam and other boilers (711)	98	39	59	−0.275	4.338
Steam turbines (712)	26	27	−1	+0.202	1.565
Internal combustion engines (713)	805	898	−93	+0.227	2.073
Engines and motors (714)	63	78	−15	+0.274	0.102
Rotating electric plant and motors (716)	341	566	−225	+0.405	2.629
Other power generating machinery (718)	51	33	18	−0.046	0.799
Specialised by industry (72)	3 573	938	2 635	−0.448	0.952
Agricultural machinery (721)	524	165	359	−0.382	1.733
Tractors (722)	87	88	−1	+0.171	1.821
Civil engineering, construction equip (723)	351	211	140	−0.080	1.540
Textile and leather machinery (724)	522	139	383	−0.449	1.050
Paper-making machinery (725)	163	28	135	−0.599	0.530
Printing, bookbinding equipment (726)	247	28	219	−0.725	0.205
Food processing equipment (727)	311	31	280	−0.753	0.978
Other specialised equipment (728)	1 367	269	1 098	−0.563	0.721
Metal working machinery (73)	809	384	425	−0.193	0.899
Machine tools for removing metal (731)	312	144	168	−0.208	0.585
Machine tools for working metal (733)	147	33	114	−0.512	0.814
Parts and accessories for (731, 733, 735)	122	125	−3	+0.187	1.567
Other metalworking machinery (737)	228	82	228	−0.322	1.355
General industrial machinery (74)	4 482	1 419	3 064	−0.379	0.893
Heating and cooling equipment (741)	1 056	224	832	−0.537	0.693
Pumps for liquid (742)	333	144	189	−0.239	1.066
Pumps not for liquid , compressors (743)	676	149	527	−0.522	0.468
Mechanical handling equipment (744)	493	220	273	−0.225	1.397
Non-electrical machinery (745)	725	92	633	−0.695	0.494
Ball bearings (746)	130	169	−39	+0.301	1.629
Taps, cocks, valves (747)	571	152	417	−0.451	0.973
Transmission shafts (748)	201	146	55	+0.016	1.181
Non-electrical parts (749)	298	124	174	−0.257	1.433

Source: estimated from COMEXT database.

These were steam turbines (SITC 712), internal combustion engines (SITC 713), rotating electric plant and motors (SITC 716), tractors (SITC 722) parts and accessories for metal-working machine tools (SITC 735) and transmission shafts (SITC 748).Quantitatively, the most important items for which a revealed specialisation was observed were internal combustion engines and rotating electric plant which were priority sectors under central planning. Production and trade in general industrial equipment was subject to a greater degree of concentration by exporting-country than many other industrial sectors, with the result that several CEE economies recorded a revealed specialisation in specific SITC three-digit categories where the CEE-10 as a whole failed to do so (see Table 7.7). Exports from the Baltic states were negligible.

The CEE-10 recorded a revealed specialisation in power generating equipment (SITC 71) which resulted from the strong export perform-ance of Hungary and the Czech Republic. Hungary had become the

Table 7.7   CEE-10 economies with revealed specialisation in machinery and equipment for general industrial purposes by country

SITC category and number	Countries recording positive RCA index and ESI > 1, ranked by ESI
Steam and other boilers (711)	Sla (11.9) Est (11.5) Hun (3.3) Bul (1.9)
Steam turbines (712)	Pol (2.7) Hun (1.8) Cz (1.7) Sle (1.2)
Internal combustion engines (713)	Hun (9.6)
Rotating electric plant and motors (716)	Hun (2.9) Cz (4.3) Sle (4.3) Sla (3.1) Rom (2.4) Bul (1.6)
Other power generating machinery (718)	Bul (5.3) Cz (1.1)
Agricultural machinery (721)	Hun (3.8) Est (1.5)
Tractors (722)	Cz (3.8) Pol (2.7) Sla (2.2),
Construction equipment (723)	Sle (2.6) Hun (1.7) Sla (1.7)
Textile and leather machinery (724)	Est (2.0)
Paper-making machinery (725)	Est (1.0)
Machine tools for removing metal (731)	Cz (1.7) Bul (1.3)
Parts and accessories for 731 (733,735)	Cz (3.3) Sla (2.2) Sle (1.7) Rom (1.3)
Other metalworking machinery (737)	Bul (2.7)
Pumps for liquid (742)	Cz (2.6)
Pumps not for liquid , compressors (743)	Sle (1.4)
Mechanical handling equipment (744)	Bul (4.0)
Ball bearings (746)	Rom (6.2) Sla (3.5) Pol (1.9)
Taps, cocks, valves (747)	Sle (1.7)
Transmission shafts (748)	Cz (2.3) Lat (2.3) Sle (1.3) Pol (1.0)

Source: estimated from COMEXT database.

fourth largest exporter of power generating equipment to the EU in the world (following the USA, Japan and Switzerland) with exports of ECU 957 million which accounted for 55.4 per cent of CEE-10 exports in this category. These had grown from only ECU 32 million in 1990. Hungarian exports of internal combustion engines (SITC 713) of ECU 767 million were the largest component and accounted for 85.3 per cent of CEE-10 exports in this sector. The Czech republic was the eighth largest world exporter of power generating equipment to the EU with exports of ECU 255 million. Bulgaria's revealed specialisation in mechanical handling equipment (SITC 744), which was the legacy of its specialisation in the production of fork-lift trucks in the communist period was based on a low volume of exports (ECU 22 million compared with ECU 17 million in 1988). This reflected a failure to divert exports to EU markets to compensate for the loss of the Soviet market which was the major market for these products in the communist period.

7.3.3 Trade in equipment for offices and telecommunications equipment and electrical equipment

The EU-15 recorded a surplus of ECU 3798 million in its trade with the CEE-10 in office, telecommunications and electrical equipment (SITC 75–7) in 1996. This was an important area where the pattern of EU trade with the CEE-10 differed substantially from the general pattern of surpluses and deficits in extra-EU trade as a whole. The EU recorded large deficits in its external trade in office machinery and computers (SITC 75) in all years in the early 1990s. The EU-15 deficit reached ECU 20,687 million in 1996 as a result of substantial deficits in trade in computer equipment with the USA (ECU 3803 million) Japan (ECU 2691 million), Singapore (ECU 2803 million) and smaller deficits in trade with Malaysia, Taiwan and South Korea. However, the CEE-10 accounted for only 1.2 per cent of extra-EU imports of office machines and computers and only 0.89 per cent of extra-EU imports of computers in 1996. Although Hungarian exports of computers of ECU 134 million generated a surplus of ECU 44 million, this was offset by a deficit of ECU 117 million on trade in computer spare parts and accessories. Computers represented 30 per cent of Bulgarian exports of machinery and equipment to the Soviet Union in the late 1980s and the inability to penetrate EU markets for computers has created major problems for the Bulgarian economy. Poland was the seventh largest market for EU computers in 1996 (ECU 211 million) and the Czech Republic the tenth largest (ECU 183 million).

The EU-15 also recorded deficits in trade in telecommunications equipment (SITC 76) in total extra-EU trade in the early 1990s and only moved into surplus in this category (ECU 632 million) in 1996, as a result of its surplus in trade with the CEE-10 of ECU 1298 million. The EU surplus in trade with the CEE-10 was the result of the high level of demand for telecommunications equipment (SITC 764) from Poland, Hungary and the Czech Republic to modernise their telephone and communications network . EU trade with the rest of the world in electrical machinery (SITC 77) also moved into surplus (ECU 1624 million) in 1996 from deficits in 1994 (ECU 2229 million) and 1995 (ECU 2800 million) partly as a result of the surplus in trade with the CEE-10 of ECU 1531 million in 1996.

The only sector in which the combined CEE-10 recorded a revealed specialisation was SITC 773 (equipment for distributing electricity) which predominantly consists of insulated wires and cables and

Table 7.8   EU-15 trade with the CEE-10 in machinery and equipment for office, telecommunications and electrical machinery in 1996 (ECU million)

	EU export	EU import	Balance	RCA	ESI
Total (75–7)	8 583	4 785	3 798		
Office machines and computers (75)	1 419	450	969	−0.378	0.126
Office machines (751)	135	16	119	−0.715	0.093
Computers (752)	686	184	502	−0.448	0.093
Parts and accessories for 751–2 (759)	597	250	347	−0.252	0.178
Telecommunications & recording (76)	2 075	777	1 298	−0.305	0.410
Television receivers (761)	108	231	−123	+0.504	0.693
Radio receivers (762)	66	25	41	−0.289	0.099
Sound recording apparatus (763)	72	113	−41	+0.384	0.436
Telecommunications equipment (764)	1 830	407	1 423	−0.519	0.319
Electrical machinery (77)	5 089	3 558	1 531	−0.003	0.874
Electric power machinery (771)	412	267	145	−0.041	**1.001**
Electrical apparatus related to circuits (772)	1 315	608	707	−0.207	**1.004**
Equipment for distributing electricity (773)	903	1 071	−168	+0.256	**3.485**
Electro-diagnostic apparatus (774)	145	12	132	−0.782	0.089
Household electrical equipment (775)	809	545	264	−0.021	**2.342**
Valves and transistors (776)	474	292	182	−0.066	0.186
Electrical machinery & apparatus, batteries, filaments, lamps (778)	1 033	764	269	+0.025	0.807

Source: estimated from the COMEXT database.

insulating equipment, was one of the fastest growing sectors of CEE-10 exports to the EU in the first half of the 1990s. All CEE-10 countries, except Bulgaria, Slovenia and Latvia, recorded a revealed specialisation in this sector. Exports from the CMEA-5 to the EU-12 which stood at only ECU 43 million in 1988 had grown to ECU 435 million by 1994, with particularly rapid growth recorded by Hungary, Slovakia, the Czech Republic, and to a lesser extent Poland. Although the power distribution sector was a relatively important area of intra-CMEA cooperation in the Soviet era, there is no statistical evidence to show that the growth of exports from Hungary, the Czech Republic and Slovakia to the EU was the result of the re-orientation of intra-CMEA trade to the EU which could be taken to indicate the successful exploitation of a competitive advantage inherited from the Soviet era. Soviet, Hungarian and Czechoslovak trade data for the late 1980s all indicate that Soviet imports of wires, cables and insulating material from Hungary and Czechoslovakia were relatively small.

Table 7.9 indicates that few countries, except Hungary, registered a revealed specialisation in the remaining three-digit sectors in office equipment and electrical machinery. Bulgaria failed to register a revealed specialisation in any sector whatsoever, despite the predominance of exports of computers and office equipment in its exports to the USSR in the Soviet period. CEE export performance in electrical consumer goods and household electrical equipment was poor, with the majority

Table 7.9 CEE-10 economies with revealed specialisation in machinery and equipment for office, telecommunications and electrical machinery in 1996

SITC number and description	Countries recording positive RCA index and ESI > 1, ranked by ESI
Parts for computers and office equipment (759)	Est (1.9)
Television receivers (761)	Hun (6.7) Pol (3.0) Sle (2.3)
Sound recording apparatus (763)	Hun (2.12)
Electric power machinery (771)	Est (4.5) Lat (2.9) Hun (1.4)
Equipment for distributing electricity (773)	Hun (6.8) Sla (6.4) Lit (5.3) Cz (2.9) Pol (2.8) Est (2.4) Rom (2.0)
Household electrical equipment (775)	Sle (10.0) Hun (2.9)
Valves and transistors (776)	Lit (1.4)
Electrical machinery & apparatus, batteries, filaments, lamps (778)	Hun (1.8)

Source: estimated from the COMEXT database.

of countries recording a negative RCA index and an ESI of less than one in the majority of three-digit categories. Hungary, Poland and Slovenia registered a revealed specialisation in the production of TV sets and Slovenia in household electrical equipment.

Hungary's performance in this sector is exceptional with a revealed specialisation in six of the fourteen three-digit categories. Details of Hungary's trade with the EU-15 in SITC 75–7 are shown in Table 7.10. Hungarian exports of office, telecommunications and electrical equipment to the EU-12 grew from ECU 117 million in 1988 to ECU 672 million in 1994. The fastest growing sectors were cables and insulating materials (SITC 773), household electrical equipment and miscellaneous electrical machinery and apparatus (SITC 718). Hungary also recorded significant export growth in two categories in which it has a large trade deficit with the EU, namely, telecommunications equipment (SITC 764) and electrical apparatus linked to circuits (SITC 772). However, Hungary

Table 7.10   EU-15 trade with Hungary in machinery and equipment for office, telecommunications and electrical machinery in 1996 (ECU million)

	EU export	EU import	Balance	RCA	ESI
Total (75–7)	1 944	1 822	122		
Office machines and computers (75)	350	252	98	−0.036	0.385
Office machines (751)	25	1	24	−0.866	0.044
Computers (752)	91	134	−43	+0.310	0.367
Parts and accessories for 751–2 (759)	234	117	117	−0.217	0.451
Telecommunications & recording (76)	483	448	35	+0.087	1.286
Television receivers (761)	9	124	−115	+0.890	6.737
Radio receivers (762)	23	22	1	+0.100	0.487
Sound recording apparatus (763)	6	102	− 96	−0.916	2.124
Telecommunications equipment (764)	445	200	225	−0.268	0.852
Electrical machinery (77)	1 111	1 122	− 11	+0.130	1.499
Electric power machinery (771)	79	69	10	+0.060	1.415
Electrical apparatus related to circuits (772)	286	181	105	−0.102	**1.623**
Equipment for distributing electricity (773)	275	387	−112	+0.287	**6.835**
Electro-diagnostic apparatus (774)	17	5	12	−0.424	0.204
Household electrical equipment (775)	77	125	− 48	+0.355	**2.927**
Valves and transistors (776)	162	42	120	−0.497	0.147
Electrical machinery & apparatus, batteries, filaments, lamps (778)	215	313	− 98	+0.303	**1.796**

Source: estimated from the COMEXT database.

has been unsuccessful in penetrating EU markets for electrical apparatus for medical diagnostic purposes although medical instruments were an important source of exports to the Soviet Union.

7.3.4 Trade in transport equipment

The EU-15 made a surplus of ECU 3161 million in trade in transport equipment with the CEE-10 in 1996 as a result of a surplus in trade in road vehicles (SITC 78) of ECU 3188 million. Nevertheless, the CEE-10 recorded a revealed specialisation in 3 of the 9 three-digit categories under this heading, namely, passenger cars (SITC 781), trailers and caravans (SITC 786) and railway vehicles and locomotives (SITC 791). ESI indices of greater than one combined with negative RCA indices were recorded by the combined CEE-10 in 3 of the 4 three-digit sectors of road vehicles. This reflected the relative lack of success in penetrating the EU market for road vehicles other than passenger cars, achieved by manufacturers outside the EU.

The CEE-10 revealed specialisation in road transport vehicles can be largely attributed to exports from central-eastern Europe, while the Balkan and Baltic states (with the exception of Lithuania which exported motorbikes and cycles) had made little progress in EU markets for road

Table 7.11 EU-15 trade with the CEE-10 in machinery and equipment for transport in 1996 (ECU million)

	EU export	EU import	Balance	RCA	ESI
Total (78–9)	7 160	3 999	3 161		
Road vehicles (78)	6 826	3 639	3 188	−0.138	1.565
Passenger cars (781)	2 989	2 113	876	+0.003	1.706
Vehicles for transport of goods and special purpose vehicles(782)	807	203	604	−0.473	1.423
Other road vehicles (783)	393	16	377	−0.890	1.501
Parts and accessories for road vehicles (784)	2 244	917	1 327	−0.265	1.614
Motorbikes and cycles (785)	116	111	5	+0.150	0.348
Trailers, caravans (786)	277	279	− 2	+0.179	5.990
Other transport equipment (79)	334	360	− 26	+0.211	0.210
Railway vehicles, locomotives (791)	182	132	50	+0.018	3.777
Aircraft and associated equipment (792)	98	28	70	−0.428	0.020
Ships and boats (793)	53	200	−147	+0.685	0.726

Source: estimated from the COMEXT database.

vehicles by 1996. The Czech Republic, Slovenia and Slovakia, all recorded an unambiguous competitive advantage in the production of passenger cars (SITC 781). Poland, which is the fourth largest extra-EU exporter of cars to the EU, recorded a negative RCA index in this sector in 1996 as a result of a major increase in imports of cars from the EU-15 from ECU 561 million in 1995 to ECU 1129 million in 1996 despite recording an ESI of 1.9. Table 5.11 shows that exports of cars from central-east Europe (Poland, Slovenia, the Czech Republic, Slovakia and Hungary which were the fourth to eighth largest exporters of cars to the EU respectively in 1996) grew by 129 per cent between 1993 and 1998 to ECU 2103 million and constituted 16.4 per cent of EU-15 imports of passenger cars from outside the EU in 1996 compared with 7.5 per cent in 1993, largely displacing imports from Japan. CEE-10 exports of passenger cars to the EU were directed at below average quality markets. The average unit value of exports of passenger cars from the CEE-10 to the EU was 86.5 per cent of the average EU import price and 72.2 per cent of the unit value of passenger cars sold in intra-EU trade. Poland's high volume of car exports were achieved at a unit value of 52.9 per cent of the unit value for intra-EU trade and the Czech Republic's achieved 62.7 per cent. However Slovenia and Slovakia both succeeded in selling substantial volumes of cars at unit values that were 1.9 per cent and 19.5 per cent respectively above the average for intra-EU trade.

CEE-10 surpluses in trade with the EU-15 in transport equipment other than road vehicles (SITC 79) in 1996 resulted from exports of

Table 7.12   CEE-10 Economies with revealed specialisation in trade in transport equipment with EU-15 in 1996

SITC number and description	Countries recording positive RCA index and ESI > 1, ranked by ESI
Passenger cars (781)	Sle (4.1) Sla (2.8) Cz (1.8)
Special purpose vehicles (782)	Pol (4.2)
Parts and accessories for road vehicles (784)	Cz (2.6)
Motorbikes and cycles (785)	Lit (7.8)
Trailers, caravans (786)	Hun (8.0) Pol (7.9) Sla (6.3) Sle (6.1) Cz (4.8)
Railway vehicles, locomotives (791)	Sla (11.7) Cz (8.6)
Ships and boats (793)	Pol (2.1) Lit (1.7) Rom (1.1)

Source: estimated from the COMEXT database.

Table 7.13   EU-15 trade in passenger cars (ECU million)

	EU imports			EU exports	EU balance
	1993	1995	1996	1996	1996
Total	12 254	11 624	12 853	33 935	+21 082
Japan	8 577	6 738	7 185	5 802	−1 383
USA	1 199	1 133	1 377	9 147	+7 770
Poland	434	526	607	1 129	+522
Slovenia	213	453	534	438	−96
Czech	236	319	466	668	+202
Slovakia	23	191	272	176	−96
Hungary	13	140	224	330	+106
Total Central Europe	919	1 103	2 103	2 741	+638

Source: estimated from the COMEXT database.

railway locomotives (SITC 791) from the Czech Republic and Slovakia and of ships and boats (SITC 793) from Poland, which was the ninth largest exporter of ships to the EU in that year, with exports of ECU 144 million;[2] Romania (ECU 25 million) was the only other significant exporter of ships and boats to the EU with Lithuania's competitive advantage being recorded on a very low volume of exports (ECU 8 million).

7.4   Comparative advantage and revealed specialisation in trade in miscellaneous manufactured goods

7.4.1   General patterns of trade in miscellaneous manufactured goods

Trade between the EU and the CEE-10 in miscellaneous manufactured goods classified by materials (SITC 6) and miscellaneous manufactured articles (SITC 8) are heavily affected by outward processing arrangements. EU exports of goods for outward processing often appear under SITC 6 and contribute to EU imports under SITC 8. Consequently the two categories have been amalgamated in Table 7.14 to analyse the net factor content of trade.

The EU was a net exporter of ECU 1322 million in trade in human capital-intensive goods and a net importers of labour-intensive goods of ECU 1243 million and resource-intensive goods of ECU 1249 million in trade with the CEE-10 (see Table 7.14).

Table 7.14   Factor content of EU trade with CEE-10 in miscellaneous manufactured goods and equipment

	EU exports to CEE-10	EU imports from CEE-10	Balance
Human capital-intensive	3 046	1 724	+1 322
High technology	836	204	+622
Medium technology	1 916	1 252	+464
Low technology	304	68	+236
Labour-intensive	12 470	13 713	−1 243
Resource-intensive	5 840	7 094	−1 249
Total	21 361	22 531	−1 170

Source: estimated from the COMEXT database.

7.4.2   Trade in manufactures of leather, rubber, wood, paper and textiles (SITC 60–5)

The combined CEE-10 recorded a revealed specialisation in 6 of the 20 three-digit sectors in this category, namely leather manufactures (SITC 612), tyres and inner tubes (SITC 625), veneers, plywood and particle board (SITC 634), wood manufactures (SITC 635) textile yarns (SITC 651) and made-up articles of textile materials (SITC 658). The strongest specialisation was revealed in wood manufactures where the combined CEE-10 recorded an RCA index of 0.832 and an ESI of 4.207. All of the CEE-10 economies, except Romania and Bulgaria had an individual revealed specialisation in wood manufactures in a sector where the CEE-10 accounted for 40.5 per cent of EU-15 imports from outside the EU. The majority of EU exports of textile yarns and fibres (57 per cent) were exported under outward processing agreements. When outward processing trade is removed from both EU exports and imports in SITC 65, the EU surplus falls from ECU 3092 million to ECU 830 million.

The EU-15 also made a substantial surplus in trade in paper and paperboard and related products (SITC 64) of ECU 1221 million and a small surplus in trade in leather (SITC 611) of ECU 449 million. The EU surplus in trade in paper and paperboard and related products reflects the low priority attached to the consumption and production of commodities for the packaging, publishing and data processing industries under communism. Consequently, the sudden growth in demand for these products folowing the liberalisation of the press, has resulted in the growth of EU exports to the former CMEA economies from only ECU 123 million in 1988 to ECU 1594 million in 1996. There are indications of

Table 7.15   EU-15 trade with the CEE-10 in manufactures of leather, rubber, wood, paper and textile yarns in 1996 (SITC 61–5) (ECU million)

	EU export	EU import	Balance	RCA	ESI
Miscellaneous manufactures (6)	13 783	11 290	2 493	0.076	1.549
SITC 60–5	7 922	3 995	3 927	−0.164	1.460
Leather leather manufactures and dressed furs (61)	707	228	479	−0.372	0.985
Leather (611)	624	175	449	−0.431	0.895
Leather manufactures (612)	13	18	−5	0.331	1.075
Furskins (613)	70	35	35	−0.169	1.840
Rubber manufactures (62)	479	569	−90	0.256	1.634
Rubber materials (621)	153	80	73	−0.144	1.666
Tyres and inner tubes (625)	138	379	−241	0.594	1.728
Rubber articles (629)	188	109	79	−0.097	1.356
Cork and wood manufactures (63)	320	1 116	−796	0.665	3.046
Cork manufactures (633)	29	1	28	−0.987	0.068
Veneers plywood, particle board (634)	187	317	−130	0.413	1.815
Wood manufactures (635)	104	799	−695	0.832	4.207
Paper, paperboard and articles (64)	1866	645	1 221	−0.341	1.271
Paper and paperboard	1 189	415	774	−0.336	1.105
Articles of paper and paperboard	677	230	447	−0.348	1.749
Textile yarn and fabrics (65)	4 529	1 437	3 092	−0.378	1.127
Textile yarn (651)	394	366	28	0.138	1.219
Cotton fabrics, woven (652)	695	241	455	−0.340	1.386
Fabrics of man-made textiles, woven (653)	1 288	152	1 136	−0.712	0.750
Other textile fabrics, woven (654)	612	82	530	−0.679	1.871
Knitted or crocheted fabrics (655)	413	30	383	−0.813	0.681
Tulles, lace, embroidery, trimmings (656)	138	23	115	−0.619	0.698
Special yarns and fabrics (657)	689	74	615	−0.735	0.695
Made-up articles of textile materials (658)	112	440	− 328	0.696	1.856
Floor coverings and carpets (659)	187	29	159	−0.644	0.215

Source: estimated from the COMEXT database.

(vertical) intra-industry trade developing in this area reflected in the ESI indices of greater than one and a number of CEE economies taking a growing share of the EU market. Poland was the fifth largest world exporter of paper and paperboard products to the EU in 1996, followed by Slovenia in eighth place, the Czech Republic in tenth and Slovakia in eleventh.

Table 7.16   CEE-10 economies with revealed specialisation in manufactures of leather, rubber, wood, paper and textiles yarns in 1996 (SITC 61–5)

SITC number and description	Countries recording positive RCA index and ESI > 1, ranked by ESI
Leather (611)	Lit (4.5) Lat (1.2)
Leather manufactures (612)	Lat (3.1) Pol (2.1) Est (1.1)
Furskins (613)	Est (8.1) Pol (4.9) Sle (1.6)
Rubber materials (621)	Sla (3.1) Bul (2.1)
Tyres and inner tubes (625)	Sle (3.3) Sla (2.0) Pol (1.8) Hun (1.3) Rom (1.1)
Rubber articles (629)	Cz (2.3), Rom (1.3) Bul (1.3)
Veneers, plywood, particle board (634)	Lat (29.0) Est (8.1) Lit (4.6) Bul (2.4) Pol (2.1) Sle (1.7) Cz (1.3)
Wood manufactures (635)	Lat (10.1) Pol (7.5) Est (6.6) Sle (6.2) Lit (4.2) Cz (3.2) Hun (2.0) Sla (1.9)
Paper and paperboard (641)	Sla (2.7) Sle (2.1)
Articles of paper and paperboard (642)	Sle (3.9) Sla (2.7)
Textile yarn (651)	Lat (7.2) Sla (3.9) Lit (2.9) Sle (2.0)
Cotton fabrics, woven (652)	Est (10.9) Lat (4.8) Lit (2.9) Cz (2.8)
Other textile fabrics, woven (654)	Lat (8.8) Cz (3.8)
Made-up articles of textile materials (658)	Est (5.3) Lat (5.1) Lit (2.9) Pol (2.5) Cz (2.0) Bul (1.7) Hun (1.5) Sle (1.2) Rom (1.0)

Source: estimated from the COMEXT database.

CEE economies with a revealed specialisation in trade in three-digit SITC categories under SITC 61–5 are shown in Table 7.16. The Baltic states featured prominently in the majority of sectors, with revealed specialisations and relatively high ESI indices in the production of wood products, leather and made-up textile articles which include bed and table linens. Re-imports under outward processing arrangements account for 40 per cent of EU imports from the CEE-10 under this heading.

7.4.3 Trade in mineral and metal manufactures

Exports of minerals and metal manufactures were a significant component of CEE-10 exports to the EU-15 accounting for 18.8 per cent of EU-15 imports of manufactured goods from the CEE-10 and for 15.3 per cent of EU-15 imports of these goods from outside the EU. Outward processing trade was relatively unimportant in this sector. The combined CEE-10 recorded a surplus in trade in minerals and metals with the EU-15 of ECU 1433 million in 1996 and a revealed specialisation in 19 of the 33 SITC three-digit sectors (see Table 7.17).

Table 7.17 EU-15 trade with the CEE-10 in mineral and metal manufactures in 1996 (SITC 66–9) (ECU million)

	EU export	EU import	Balance	RCA	ESI
Minerals and metals (66–9)	5 861	7 294	−1 433	0.279	1.229
Non-metallic mineral manufactures (66)	1 235	1 392	−157	0.232	0.985
Lime, cement, construction materials (661)	144	380	−236	0.578	5.864
Clay construction materials (662)	381	118	263	−0.389	3.402
Mineral manufactures (663)	274	204	70	0.029	2.129
Glass (664)	235	228	7	0.159	2.600
Glassware (665)	148	298	−150	0.483	3.884
Pottery (666)	46	164	−118	0.668	2.307
Precious metals, stones, pearls (667)	7	1	6	−0.768	0.001
Iron and Steel (67)	1 550	2 216	−666	0.341	2.737
Pig iron, granules and ferro-alloys (671)	29	203	−174	0.817	0.875
Ingots, primary and semi-finished products of iron and steel (672)	25	59	−35	0.546	1.068
Flat-rolled products, non-alloy, unclad (673)	188	851	−663	0.731	4.940
Flat-rolled products, non-alloy, clad (674)	244	43	201	−0.602	0.944
Flat-rolled products of alloy steel (675)	273	89	184	−0.367	2.278
Iron and steel bars, rods, shapes (676)	240	492	−252	0.489	4.644
Rails and railway track (677)	34	36	−2	0.200	7.901
Wire of iron and steel (678)	63	79	−16	0.282	3.294
Tubes, pipes, hollow fittings (679)	454	364	90	0.066	2.776
Non-ferrous metals (68)	730	1 540	−810	0.500	1.172
Silver and platinum (681)	19	81	−62	0.722	0.587
Copper (682)	185	642	−457	0.663	1.547
Nickel (683)	39	14	25	−0.329	0.122
Aluminium (684)	382	678	−296	0.433	1.394
Lead (685)	17	43	−26	0.565	1.552
Zinc (686)	33	66	−33	0.475	2.585
Tin (687)	4	1	3	−0.729	0.016
Miscellaneous non-ferrous metals (689)	51	15	36	−0.400	0.187
Manufactures of metals (69)	2 346	2 146	200	0.131	1.549
Structures and parts of structures of iron, steel and aluminium (691)	396	541	−145	0.320	5.772
Metal containers for storage & trnsprt (692)	214	142	72	−0.027	3.679
Wire products, fencing (693)	75	65	10	0.098	2.452
Nails, screws, nuts, bolts etc. (694)	147	140	7	0.151	1.039
Tools for use by hand or machine (695)	285	135	150	−0.195	0.657

Table 7.17   *(Contd)*

	EU export	EU import	Balance	RCA	ESI
Cutlery (696)	46	16	30	−0.347	0.256
Household goods made of base metal (697)	228	151	77	−0.030	**1.257**
Manufactures of base metal (699)	956	958	−2	0.175	**2.796**

Source: estimated from the COMEXT database.

Table 7.18   CEE-10 economies with revealed specialisation in trade in minerals and metals in 1996 (SITC 66–9)

SITC number and description	Countries recording positive RCA index and ESI > 1 , ranked by ESI
Lime, cement, construction materials (661)	Lit (10.6) Sla (10.5) Pol (9.6) Est (6.7) Cz (6.4) Lat (6.2) Rom (5.8) Bul (2.3)
Clay construction materials (662)	Cz (5.9)
Mineral manufactures (663)	Sle (4.7) Czech (3.4) Pol (2.0)
Glass (664)	Cz (5.6) Lit (2.9) Hun (2.5) Sla (1.9) Lat (1.6) Rom (1.0)
Glassware (665)	Cz (6.7) Bul (5.7) Rom (4.3) Sla (3.8) Pol (3.7) Sle (2.9) Est (2.3) Hun (1.4)
Pottery (666)	Rom (4.4) Czech (3.6) Pol (3.5) Bul (1.3)
Pig iron, granules and ferro-alloys (671)	Lat (10.0) Est (4.8) Sla (2.9) Rom (1.3)
Ingots, primary and semi-finished products of iron and steel (672)	Sla (1.9) Bul (1.8) Poland (1.3) Hun (1.2) Cz (1.2)
Flat-rolled products, non-alloy, unclad (673)	Bul (24.9) Sla (14.8) Rom (12.2) Hun (4.4) Cz (2.6) Pol (1.8)
Flat-rolled products, non-alloy, clad (674)	Sla (6.3) Bul (3.8)
Flat-rolled products of alloy steel (675)	Sle (10.0)
Iron and steel bars, rods, shapes (676)	Cz (8.1) Pol (6.6) Rom (5.2) Sle (4.1) Bul (2.0)
Rails and railway track (677)	Pol (24.6) Cz (6.8)
Wire of iron and steel (678)	Lat (10.7) Cz (9.4) Sla (4.1) Bul (3.1) Lit (2.5) Rom (2.2)
Tubes, pipes, hollow fittings (679)	Rom (5.0) Cz (4.4) Sla (4.2) Pol (2.4)
Silver and platinum (681)	Pol (2.2)
Copper (682)	Bul (7.5) Pol (4.2)
Aluminium (684)	Sla (3.5) Rom (2.8) Sle (2.2) Hun (2.1)
Lead (685)	Bul (24.2) Pol (1.8) Lat (1.1)
Zinc (686)	Pol (6.4) Bul (5.9) Rom (2.0)

Table 7.18   (*Contd*)

SITC number and description	Countries recording positive RCA index and ESI > 1 , ranked by ESI
Miscellaneous non-ferrous metals (689)	Lat (5.1) Est (2.7) Lit (2.0)
Structures and parts of structures of iron, steel and aluminium (691)	Pol (10.1) Est (9.0) Cz (6.5) Hun (4.8) Sla (4.7)
Metal containers for storage & trnsprt (692)	Cz (6.8) Sla (4.4) Pol (4.0) Lat (3.1)
Wire products, fencing (693)	Pol (3.9) Cz (3.7) Sla (2.5)
Nails, screws, nuts, bolts etc. (694)	Cz (1.6) Pol (1.2) Rom (1.2) Bul (1.2)
Tools for use by hand or machine (695)	Sle (1.3)
Household goods made of base metal (697)	Hun (2.0) Cz (1.4) Sla (1.2)
Manufactures of base metal (699)	Cz (5.0) Sle (3.1) Pol (3.0) Sla (2.1)

Source: estimated from the COMEXT database.

In 28 of the 33 sectors, at least one CEE economy recorded a revealed specialisation (see Table 7.18). Strong CEE-10 revealed specialisations (reflected in an aggregate ESI of greater than 3.8, an RCA index of greater than 0.3 and five or more CEE countries showing a revealed specialisation) were recorded in trade in lime, cement and construction materials (SITC 661), glassware (SITC 665), flat-rolled steel products, non-alloy, unclad (SITC 673), iron and steel bars, rods and shapes (SITC 676), structures of iron, steel and aluminium (SITC 691). Copper (SITC 684) and aluminium products (SITC 691) were dominant in CEE-10 exports of non-ferrous metals. Manufactures of base metal (SITC 699) were also an important source of exports for the CEE-10.

7.4.4   Trade in light consumer goods

The CEE-10 made a surplus in trade in light consumer goods (SITC 82–5) of ECU 6336 million in 1996. These items consist of labour-intensive goods, many of which are produced under outward processing arrangements which were examined in Chapter 5. The CEE-10 trade surplus in light consumer goods falls to ECU 2565 million after CEE trade conducted under outward processing agreements has been netted off from both exports and imports in categories SITC 82–5. Imports and re-exports related to outward processing trade have been included in the estimates of RCA and ESI indices for the reasons outlined in Chapter 3.

Table 7.19 EU-15 trade with the CEE-10 in light industrial consumer goods (SITC 82–5) (ECU million)

	EU export	EU import	Balance	RCA	ESI
Total (82–5)	2 879	9 215	−6 336		
Furniture and bedding, parts (821)	733	2 306	−1 573	0.634	4.670
Travel goods, handbags and cases (831)	71	112	−41	0.385	0.472
Clothing and apparel (84)	1 466	5 630	−4 164	0.691	1.741
Male clothing, not crocheted (841)	201	1 660	−1 459	0.843	2.156
Female clothing, not crocheted (842)	197	2 044	−1 847	0.873	2.622
Male clothing , crocheted (843)	57	110	−52	0.461	0.820
Female clothing, crocheted (844)	164	360	−196	0.515	1.484
Garments, predominantly made up from textile fabrics (845)	417	1 106	−689	0.581	1.217
Clothing accessories made up from textile fabrics (846)	324	218	106	−0.022	1.682
Clothing and accessories not made up from textile fabrics (848)	105	133	−28	0.287	0.491
Footwear (851)	609	1 167	−558	0.463	2.048

Source: estimated from the COMEXT database.

Table 7.20 CEE-10 economies with revealed specialisation in light industrial consumer goods (SITC 82–5)

SITC number and description	Countries recording positive RCA index and ESI > 1, ranked by ESI
Furniture and bedding, parts (821)	Pol (7.7) Rom (7.1) Est (6.2) Sle (5.4) Lat (4.0) Cz (3.5) Sla (2.8) Lit (2.6) Hun (1.8)
Travel goods, handbags and cases (831)	Bul (1.8)
Male clothing , not crocheted (841)	Rom (6.7) Lit (4.7) Bul (3.6) Lat (3.6) Est (2.7) Pol (2.3) Sla (2.2) Sle (1.3) Hun (1.3)
Female clothing, not crocheted (842)	Lit (6.2) Rom (6.0) Lat (5.2) Pol (3.8) Bul (3.6) Est (3.4) Sle (2.2) Hung (1.7) Sla (1.3)
Male clothing , crocheted (843)	Lit (2.1) Bul (2.0) Lat (1.8) Rom (1.7) Est (1.4)
Female clothing, crocheted (844)	Lit (4.2) Lat (3.4) Bul (2.3) Hun (1.8) Rom (1.7) Sla (1.4) Est (1.1)
Garments, predominantly made up from textile fabrics (845)	Lat (3.8) Rom (2.7) Lit (2.6) Bulg (2.1) Est (2.1) Hung (1.3) Pol (1.1)
Clothing accessories made up from textile fabrics (846)	Lat (6.4) Sle (2.7) Cz (2.4) Lit (1.7) Sla (1.2)
Footwear (851)	Rom (7.7) Bul (4.0) Sla (2.4) Hun (2.3) Est (2.1) Sle (1.2)

Source: estimated from the COMEXT database.

All CEE economies performed strongly in trade in light consumer goods. All ten economies recorded a revealed specialisation in several sectors of light consumer goods. Romania was the most dependent on exports in this sector recording high ESIs in exports of furniture, footwear and most categories of clothing.

7.4.5   Trade in other miscellaneous manufactures (SITC 81, 87–9)

Other miscellaneous manufactured goods, which are fall under SITC 8 include a number of items that have been evaluated as human capital-intensive. Two of these categories, optical instruments (871) and measuring and control instruments (874) have been identified as embodying high technology. CEE export performance in these sectors was poor, with a revealed specialisation for the CEE-10 combined in only two of the twenty three-digit sectors, namely prefabricated buildings (SITC 811), which was dominated by exports from Estonia, and lighting fixtures and fittings (SITC 813). The combined CEE-10 had a comparative disadvantage according to both the RCA index and the ESI in 16 of the remaining 21 categories. Slovenia was the most successful of the CEE-10 economies in this area, recording a revealed specialisation in four sectors – prefabricated buildings (SITC 811), meters and counters (873), printed matter (892) and office and stationery supplies (895).

Table 7.21   EU-15 trade with the CEE-10 in other miscellaneous manufactures (SITC 81, 87–9) (ECU million)

	EU export	EU import	Balance	RCA	ESI
Total	4 761	2 042	2 719		
Prefabricated buildings, lighting, fittings (81)	516	435	81	0.091	2.665
Prefabricated buildings (811)	47	182	−135	0.693	6.891
Sanitary, plumbing and heating fittings (812)	240	76	163	−0.378	2.239
Lighting fixtures and fittings (813)	229	177	52	0.047	1.720
Professional, scientific and controlling instruments and apparatus (87)	1 240	357	882	−0.418	0.296
Optical instruments and apparatus (871)	26	17	9	−0.051	0.231
Instruments and appliances, medical, surgical, dental and veterinary (872)	308	85	223	−0.435	0.283
Meters and counters (873)	105	68	37	−0.039	1.775

Table 7.21   (*Contd*)

	EU export	EU import	Balance	RCA	ESI
Measuring and control instruments (874)	800	187	613	−0.501	0.235
Photographic apparatus, clocks, optical goods, clocks and watches (88)	372	105	267	−0.428	0.126
Photographic apparatus (881)	59	15	44	−0.481	0.082
Photographic and cinematic supplies (882)	184	7	177	−0.898	0.038
Optical goods (884)	78	55	23	−0.001	0.367
Watches and clocks (885)	49	27	22	−0.120	0.086
Miscellaneous manufactures (89)	2 633	1 145	1 488	−0.236	0.495
Arms and ammunition (891)	19	17	2	**0.130**	0.289
Printed matter (892)	468	140	328	−0.404	0.656
Plastic articles (893)	999	410	589	−0.262	**1.129**
Toys (894)	280	269	11	**0.154**	0.425
Office and stationery supplies (895)	110	68	42	−0.069	0.773
Works of art, antiques (896)	39	16	23	−0.268	0.122
Jewellery and articles of precious metals (897)	82	27	55	−0.357	0.141
Musical instruments (898)	304	69	235	−0.515	0.213
Miscellaneous manufactures (899)	332	130	202	−0.285	0.415

Source: estimated from the COMEXT database.

Table 7.22   CEE-10 economies with revealed specialisation in other miscellaneous manufactures (SITC 81, 87–9)

SITC number and description	Countries recording positive RCA index and ESI > 1, ranked by ESI
Prefabricated buildings (811)	Est (19.9) Sle (14.2) Cz (12.7) Lit (7.3) Pol (5.3) Sla (3.9) Rom (2.7) Hun (1.5) Bul (1.1)
Sanitary, plumbing and heating fittings (812)	Hun (4.6)
Lighting fixtures and fittings (813)	Hun (2.7) Pol (2.0)
Meters and counters (873)	Sle (11.9)
Printed matter (892)	Sle (1.25)
Office and stationery supplies (895)	Sle (1.98) Cz (1.6)

Source: Estimated from the COMEXT database.

7.5 Conclusions

The more detailed sectoral analysis shows that the predominance of labour-intensive and resource-intensive goods in CEE exports to the EU did not result from the concentration of production on a small number of labour-intensive and resource-intensive sectors, but reflected a bias against exports of human capital-intensive goods across a wide range of industrial sectors. CEE exports to the EU remained concentrated on labour-intensive and resource-intensive goods in sectors in which human capital-intensive products predominated in intra-EU trade and EU imports from the rest of the world. The CEE-10 as a whole did not record a revealed specialisation in any single human capital-intensive sector embodying high technology. CEE-10 exports of chemicals were excessively biased towards resource-intensive sectors, especially fertilisers. Although the CEE-10 had positive RCA indices in trade in a number of sectors of organic chemicals, they face strong competition from exports from OECD economies. The Czech Republic, Hungary and Slovakia generally performed better than the remainder of the CEE-10 in trade in chemicals, although Bulgaria performed strongly in a few sectors. CEE export performance in machinery and equipment was poor, considering the importance of trade in machinery and equipment in the communist era, with a large deficit in trade with the EU and a low proportion of exports of goods embodying high technology. The CEE-10 revealed specialisation in machinery and equipment was concentrated on sectors of power generating equipment and equipment for distributing electricity both of which embodied low technology.

A number of sectors of machinery and equipment which featured strongly in CMEA exports to the Soviet Union, exhibit a negative RCA index combined with an ESI of greater than one in trade with the EU. These include steam boilers, agricultural machinery, civil engineering and construction equipment, electric power machinery, electrical apparatus, textile machinery, special purpose vehicles, parts and accessories for road vehicles, paper and paperboard, furs, woven cotton fabrics, other cotton fabrics, and various plastics and polymers. CEE exports in these categories commanded unit values that were substantially lower than those prevailing in EU imports from the rest of the world and intra-EU trade. The ESI of greater than one combined with negative RCAs and low unit values seems to reflect the effects of CMEA specialisation agreements which have created large capacities to produce relatively low-quality goods for which there is little demand in the EU and in which producers outside the EU have not developed production or export facilities. The long-term prospect for the development of these sectors is bleak as the progressive reduction of protective barriers to imports by the CEE economies and greater exposure to competition is likely to result in an expansion of CEE imports of these goods from the EU at the expense of domestic production.

Hungary was a notable exception to the poor export performance in machinery and equipment, recording revealed specialisations in television sets, sound-recording apparatus, household electrical equipment and higher quality electrical machinery. There were also indications of the development of export capabilities in road transport vehicles in CEE economies, especially in passenger cars, trailers and caravans. Central and east European exports of passenger cars grew rapidly largely in the mid-1990s and started to displace imports from other non-EU producers, contributing to a high ESI. The high level of domestic demand for cars in central-eastern Europe meant that some economies that were developing export capabilities in this sector (notably Poland and Hungary) failed to record an export specialisation in this sector. The gap between the unit value of EU imports of transport vehicles and imports from CEE economies was low relative to other items of machinery and equipment.

CEE revealed specialisations were heavily concentrated on industries producing miscellaneous manufactured goods falling in SITC categories 6 and 8. These included products made from leather, rubber, wood and textile materials, minerals, ferrous and non-ferrous metals. This was also reflected in high levels of revealed specialisations for products in the clothing, footwear and furniture industries. Generally, these industries involved adding relatively unskilled labour to either domestically-produced, or imported resources. Unit-values in these sectors were relatively high. The unit value of CEE exports of clothing and footwear exceeded the average unit value of CEE imports from the rest of the world. While this may reflect higher quality production, it also indicates that these industries could become vulnerable to price competition if wage rates in CEE economies start to converge on EU wage rates. All CEE economies registered significant revealed specialisations in clothing, furniture and footwear, but Romania was the most heavily dependent on exports of these items.

8
Conclusions: Prospects for Integration

8.1 The impact of enlargement

It has been shown that a major gulf exists between the economic structure and standard of living of the existing members of the EU and those of the majority of central and south-east European economies that are candidates for accession. The gap, measured in terms of income per capita, is substantially larger than that between existing members and Greece, Portugal and Spain at the time of their accession. At the same time the EU is deepening the process of economic integration between members to involve the free mobility of goods, services and factors of production and eventual monetary union. This process is expected to bring a greater concentration of industrial production, and probably services, in a limited number of regions where producers will be able to benefit from internal and external economies of scale.

Provided that the gains from trade creation EU outweigh the costs of trade diversion,[1] market enlargement will result in aggregate gains in economic efficiency for existing and new members. Existing members of the EU will benefit from static and dynamic gains in economic efficiency resulting from improved specialisation. These will include, economies of scale in production resulting from larger markets, lower-cost sources of inputs and lower cost sources of consumer goods from relatively more efficient producers inside the EU. However, it is unlikely that the efficiency gains resulting from enlargement will be distributed equally and it is possible that some regions and individuals, both in the existing EU and the accession states will become net losers from the process.

What will be the impact of enlargement on the former communist economies that are included in the first round of enlargement and those that are forced to wait for further rounds of enlargement? The nature of economic and political relations between individual CEE states and the EU remains fluid and may be substantially altered as a result of continuing tensions in former Yugoslavia, which has brought a greater political awareness of the economic difficulties facing the Balkan economies. The former communist economies can be divided into three distinct groups in terms of their future relationships with the EU, although the country-composition of the groups is by no means final. The first group will consist of those states that enter the EU in the first tier of enlargement; the second group will consists of those states that fail to be accepted in the first round of eastward enlargement but who are continuing to negotiate for accession and a third group will consist of those countries that have no firm expectation of inclusion. The last grouping currently consists of the CIS, Albania and the states that comprised former Yugoslavia (with the exception of Slovenia).

States that enter in the first tier of enlargement will benefit from guaranteed access to EU markets, net transfers from EU funds, and the possibility that they will have greater success in attracting the foreign direct investment which is required to modernise their economies. States that are not included in the first round of enlargement who have already negotiated Europe agreements will retain free trade agreements for trade in manufactured goods with the EU which will entail the full liberalisation of trade in manufactured goods. This will mean that, in principle, their exports of manufactured goods to the EU should not be greatly affected by enlargement to include other CEE economies. They will also have a longer period to prepare for compliance with the conditions established in the acquis communautaire and should, in principle, be able to use exchange-rate policy to combat fundamental balance of payments problems. In practice, politicians and businessmen fear that if their economies are not admitted in the first round of enlargement they will still be subject to measures of contingent protection, rules of origin and continued anti-dumping legislation which could affect exports of manufactured goods and could deter inward investment. Furthermore, agricultural exports from the second group will remain subject to EU controls, while controls over border movements may bring a new isolation to local economies in border regions that have developed strong cross-border links. Finally economies that are not included in the first round of enlargement will receive substantially smaller transfers from EU budgets than economies that are included in the first tier of enlargement. Financial transfers are mixed blessings and can be used to delay essential changes or can result in the displacement of domestic production by external sources. Furthermore, there are limits to the ability of the poorer economies to absorb major capital inflows. However, provided that they are used to increase the physical and human capital stock of the recipient economy they will enhance that economies productive potential over the long run. It was indicated in Chapter 5 that the economies that are not expected to be included in the first round of enlargement had been less successful in attracting capital inflows and had been required to finance imports of machinery and equipment through exports of consumer goods. This has been accompanied by increasing levels of absolute poverty (see below), cuts in real expenditure on education and reductions in school enrolment rates (Micklewright, 1999) and in falling real wages in Romania and Bulgaria. The prospect of a phased enlargement also increases the fear that FDI will be concentrated on economies that are included in the first round of enlargement and will create agglomeration effects that later entrants will be unable to imitate. There is a genuine fear that a phased enlargement will exacerbate existing income differentials between entrants in the first and subsequent rounds of enlargement and increase the economic distance between those that are included and excluded in the first round of enlargement.

The CEE economies that do enter the EU in either the first or second tier of enlargement will be required to make significant adjustment costs. New entrants will require transitional arrangements and finance to cope with the specific problems they will encounter on entry to the EU. Ultimately, full membership of the EU will require the accession states to undergo a far deeper process of economic integration than membership of a free trade area. States that are accepted for membership will be required, to fulfil the obligations of the acquis communautaire and the single market, which will involve the free circulation of goods and services, labour and capital and to comply with EU social and environmental legislation and competition law. They will also be required to enter, or at least to prepare for entry, into full monetary union. In addition they will be required to withstand competitive pressures both in EU markets and from existing EU manufacturers without recourse to devaluation.

Two further points derived from recent, and more distant, historical experience should be considered. Recent experience inside the European Union indicates that, since its accession to the EU, Spain has benefited from an inflow of foreign direct investment which has contributed to an improvement in the technological levels of Spain's exports to the EU and an increase in horizontal intra-industry trade. However, this process has also been accompanied by high levels of unemployment which are partly caused by the replacement of domestic inputs by other sources of supply within the EU which are controlled by multinational firms. More distant experience of industrialisation in the USA, under the conditions of a single market and a common currency, indicates a high degree of localisation of specific industries (which do not necessarily involve high technology) which benefit from external economies which are not available to latecomers. This implies that the removal of barriers to the mobility of goods and factors that will result from membership, and preparation for membership, will contribute to a greater concentration of production in the EU and that regions that are slow to respond to the process could be economically disadvantaged over the long term. This process could apply to the concentration of the provision of services.

This chapter will review the findings of this book and their implications for the central and south-east European economies of inclusion in, or exclusion from, this process and for EU policy towards the accession economies.

8.2   The impact of enlargement on trade flows

Empirical studies indicate that trade between high-income countries at similar levels of development tends to consist of a relatively high proportion of intra-industry trade, including horizontal intra-industry trade in similar products which are manufactured by similar production techniques. These goods are differentiated by product specifications which appeal to the tastes of specific groups of consumers who are dispersed throughout different countries. Trade between countries at different levels of economic development tends to based on differences in factor endowments. This involves greater proportions of two-way trade in completely dissimilar products and of vertically-differentiated intra-industry trade. The latter consists of trade in apparently similar goods which are differentiated by clearly defined differences in quality. Differences in quality frequently arise from differences in factor endowments involving the exchange of labour and resource-intensive goods for human capital-intensive goods. Lower-quality goods tend to be sold to lower-income consumers in the more advanced economies, while goods involving higher quality specifications are sold to relatively high income consumers in poorer economies. There is no *a priori* reason why this process should not extend to trade in services, particularly those that can be transferred and delivered electronically, with labour-intensive services concentrated in labour-intensive regions and human capital-intensive services concentrated in regions that are intensive in human capital.

Chapters 5–7 indicate that trade flows between the existing members of the EU and the majority of the accession states differ substantially from the prevailing patterns of trade between existing members of the EU and essentially involve the exchange of dissimilar goods. The central and south-east European economies are substantial net importers of human capital-intensive goods from the EU which are counterbalanced by (relative) outflows of labour-intensive and resource-intensive goods. The proportions of capital-intensive goods in EU imports from the central and east European economies are substantially below the proportions of human capital-intensive goods in trade between EU members and in EU imports from the rest of the world. This is again counterbalanced by substantially higher exports of labour-intensive and resource-intensive goods from the central and east European economies to the EU. Nevertheless, the factor content of trade flows in manufactured goods between the combined CEE-10 and the EU-15 indicates a level of trade in human capital-intensive products which is more advanced than that between Greece and the EU, that is roughly comparable with trade flows between Portugal and the EU, but substantially below that of Spain. Central and south-east European exports of human capital-intensive products to the EU lag substantially behind those of the south-east Asian economies and the EU, and are on a par with those of China, Argentina and Brazil, but are substantially higher than those of Turkey or India. The revealed specialisations of the individual countries in manufactured goods indicated that, as late as 1996, their export structures were strongly influenced by the industrial structures developed in the communist era indicating that the process of industrial restructuring will be lengthy, even in the economies that have made the greatest progress in reform.

However, the study has also demonstrated that there are major differences between the structure of EU imports from the individual CEE-10 economies. This shows that some central and south-east European economies have been more successful than others in breaking away from obsolescent industrial structures inherited from communism. Hungary had already attained a level of human capital-intensive exports that were approaching the average level of EU imports from the rest of the world by 1996 and which were on a level similar to that of Mexico. However this still remained substantially behind the level of EU imports of capital-intensive goods from the south-east Asian economies. Hungary also had the highest proportion of exports of capital-intensive goods which embodied high technology, including data processing equipment, although it did not record an export specialisation in any of these sectors. Hungary recorded a revealed specialisation in a number of medium technology sectors which have the potential for growth and development including television receivers, sound recording equipment and household electrical goods. In addition Hungary has displayed the largest improvement in export performance since the collapse of communism with half of the growth in exports attributable to human capital-intensive goods. Hungary has also made considerable progress in closing the quality gap between its exports to the EU and those of other non-EU countries, as measured by unit values. This process has involved both reductions in the gap between unit values for specific goods and a move towards a greater proportion of exports with relatively high unit values. Although the unit value of Hungarian exports of machinery and equipment was just under half that of EU imports from the world as a whole, this was the highest in the region and included a number of sectors in which the unit value exceeded the EU average. In summary, Hungary has entered a process of enhancing the technological structure of its exports to the EU and is already displaying the technical capacity to compete with producers inside and outside the EU in human capital-intensive sectors, while other sectors are showing prospects for development.

Similar conclusions can be drawn for Slovenia, and to a lesser extent, for the Czech Republic. It is not possible to make direct comparisons of changes in export structures since the collapse of communism in these two cases, as neither country existed as a separate entity in the communist era. Slovenia and the Czech Republic had the highest per capita incomes in the regions (approaching those of Greece on a purchasing power parity basis), levels of infant mortality close to the EU average (and lower than those for Greece, Portugal and the UK) and the highest levels of both male and female life expectancy in the region. Human capital-intensive goods accounted for over 40 per cent of EU imports from both Slovenia and the Czech Republic. EU imports of human capital-intensive goods from these two countries exceeded imports of labour-intensive goods. The aggregate unit value of EU imports from Slovenia was higher than from any other central and east European economy, indicating a greater ability to penetrate higher value markets in the EU.

Poland, Slovakia and Estonia occupied an intermediate position amongst the accession states, with human capital-intensive goods accounting for around 30 per cent of their exports to the EU, a proportion that was substantially below other EU economies, except Greece. In Poland's case however the relatively low share of human capital-intensive exports also reflected the relatively rapid growth of exports of clothing and furniture which were achieved in the high quality segments of the EU market. However, the unit values of exports of manufactured goods from these countries was only around a quarter of that of EU imports from the world as a whole, reflecting a relative lack of success in penetrating higher value segments of EU markets. The fact that Slovakia was not included in the first tier of negotiations for accession on the grounds that it failed to meet the political criteria established by the Copenhagen Council, rather than the economic criteria is reflected in the similarity of its economic levels with that of the other first tier economies in the intermediate category.

The economic levels of the four remaining states, all of whom are in the second tier of states negotiating accession, lag substantially behind the EU as a whole and behind the other central and east European economies. Per-capita incomes, estimated on a purchasing power parity basis for each of these economies were estimated at under ECU 4000 in 1997, which was less than a quarter of the EU average and a third of the level of Greece, the poorest of the existing members of the EU. Human capital-intensive exports accounted for just over 20 per cent of the exports of Bulgaria and Lithuania, a level that was comparable with that of Turkey, Greece and India. Romania and Latvia fell below these levels with labour-intensive goods accounting for approximately half of exports, while the ratio of human capital-intensive exports to labour-intensive exports in these two countries was comparable with India and below that of China, Turkey and Greece. Romania has had the lowest rate of growth of exports of manufactured goods to the EU of the CMEA-5 since the collapse of communism, the lowest rate of growth attributable to human capital-intensive exports and the highest growth of exports attributable to labour-intensive good, reflected in the highest proportion of exports dependent on outward processing agreements. The unit value of manufactured exports from Bulgaria, Latvia and Lithuania were the lowest of the central and east European economies, while the unit value of aggregate Romanian exports was boosted by the imported component of goods produced for re-export. The revealed specialisations of these economies were heavily concentrated in light industrial consumer goods (furniture, footwear and clothing) and resource-intensive products including basic iron and steel products, non ferrous metals, glass, pottery and, in the case of the Baltic states, leather, furs and wood products.

8.3   The role of FDI in boosting export performance

Differences in performance in EU markets reflect a combination of interrelated structural factors including the different output structures and levels of development inherited from communism, differences in the pace of reform and industrial restructuring, and differences between the levels of private sector investment and foreign direct investment in particular. The EBRD Transition Report for 1998 (pp. 82–3) indicates that there is a strong link between inflows of FDI and export performance, as FDI has contributed to both the quantity and quality of investment. FDI is a major vehicle of technology transfer, while foreign owners tend to restructure firms more quickly than domestic owners, provide more training and raise levels of labour productivity.

The importance of FDI in the process of restructuring and improving the technological structure of exports at the macroeconomic level can be illustrated by relating the data in Table 1.6 to the results found in Chapters 6 and 7. Hungary, which has made the greatest progress in upgrading its exports, has also benefited from the highest level of cumulative investment per-capita and the highest proportion of FDI in gross fixed capital formation in the region. Slovenia which has the second best export structure also had the second best *stock* of FDI per capita in 1997, (much of which predated the break-up of Yugoslavia) . Hunya (1998) has demonstrated the importance of FDI in stimulating exports at the microeconomic level. Hunya estimates that foreign-owned firms were responsible for 82.5 per cent of investment in Hungarian manufacturing industry in 1996 and for 77.5 per cent of Hungarian exports of manufactured goods. The corresponding figures for Slovenia were 20.3 per cent and 25.8 per cent. The Czech Republic and Estonia have attracted intermediate levels of FDI on a per capita basis which have made a significant contribution to gross fixed capital formation. Hunya (1998) estimates that foreign-owned firms were responsible for 33.5 per cent of investment in the Czech republic in 1996. Poland has also attracted substantial inflows of FDI since the end of 1995 which were too late to influence the estimates of export structures in 1996. Latvia and Lithuania have been relatively successful in attracting FDI from the Nordic states.

However levels of FDI in central and south-east Europe are still modest by international standards. The World Investment Report (1998, 279–83) indicates that the stock of FDI per capita in 1996 exceeded the world average in only Hungary, Estonia, the Czech Republic and Slovenia and was substantially below the average for developing countries in

Romania, Bulgaria and Slovakia. However low stock levels partly reflect the low levels of FDI in the communist era. FDI inflows, as a percentage of gross fixed capital formation between 1994 and 1996, comfortably exceeded the world average of 5 per cent in Hungary, Estonia, Poland and the Czech Republic and were close to the world average elsewhere.

The figures above lend extra credibility to the argument that there are virtuous and vicious cycles of reform, whereby economies that move quickly to implement the appropriate economic and legislative structures succeed in attracting FDI which turn provides a further impetus to reform. The EBRD quantifies the progress of reforms each year under a number of different headings to arrive at a composite index of reform. In 1998, Hungary had the highest transition index, followed by Poland, the Czech Republic, Estonia, Slovakia and Slovenia. Latvia and Lithuania had made intermediate progress in reform while the progress of Romania and Bulgaria lagged substantially behind the other accession states, despite the election of centre-right governments in 1996 (EBRD, 1998, 29). Hunya (1997) indicates that inflows of FDI are positively correlated with the speed of reform. The fast-track reformers have had greater success in attracting FDI while foreign investors have created pressures to maintain and accelerate reforms in the face of political pressures to slowdown the reform process. Foreign-owned companies have generated positive externalities in the form of imposing higher quality standards on suppliers and in providing training to employees which may subsequently be used in other employment.

8.4 Wages, productivity and poverty in central and south-east Europe

In the short-term, the price competitiveness and profitability of CEE exports to the EU depends on a combination of factors which include wage rates, levels of labour productivity and exchange rates. Labour intensive goods accounted for 40.8 per cent of EU imports from the central and south-east European economies in 1996 with levels of over 60 per cent from Romania, and around 50 per cent from the Baltic states. Central and south-east European exports of labour-intensive goods, and light consumer goods such as clothing, footwear and furniture which are produced under outward processing arrangements face severe international competition from exports from relatively low-wage economies. Low-wage economies that are dependent on labour-intensive exports are, in general, characterised by major differences between levels of income derived from estimates based on purchasing power parity and measures of income based on exchange rates. This increases the domestic purchasing power of wages while maintaining low labour costs in relation to world market prices. Although disparities exist between measures of income based on purchasing power parity and incomes measured by exchange rates within the EU in 1997 these are not substantial. It would be expected that the discrepancy between exchange-rate based measures of income and measures based on purchasing power parity would be progressively reduced as the economies become more integrated with one another and as they approach monetary union.

This section will examine the extent to which the central and south-east European economies depend on low wage rates to preserve price competitiveness and the impact of wage levels on absolute levels of poverty. Some basic data relating to these variables are shown in Table 8.1. The absolute level of poverty in central and south-east Europe has been measured by Milanovic (1998) as an income of $4 per capita per day on a purchasing power parity basis. This is a generous measure by third world standards, where the World Bank normally uses a measure of $1 per day. The $4 a day measure was equivalent to ECU 100 per month in 1997. A net wage of ECU 200 per month on a purchasing power parity basis would just permit a worker to support one (equivalent-adult) dependent on the poverty threshold, in the absence of any other form of income or welfare payment such as child benefit, maternity benefit, pension or income support.[2]

In practice the proportion of the population falling into absolute poverty at any given level of GDP will depend on a range of factors, including the distribution of income and wages and the efficiency of the welfare system. Nevertheless, the size of welfare benefits including unemployment benefit and pensions tends to be linked to average wages, while the ability to finance welfare benefits out of taxation is linked to wage levels. Consequently, the lower the average wage, the larger the percentage of the population we would expect to fall into absolute poverty. Data in Table 8.1 have been presented in country order according to average per capita monthly wage rates in manufacturing industry in 1997, expressed in ECU, which are shown in column 1. These have been converted into purchasing power parity wage rates in column 2 by using the ratio between incomes measured on an exchange rate basis and on a purchasing power parity wage basis shown in column 6. Estimates of the proportion of the population falling into poverty in column 7 have been taken from Milanovic (1998) and are based on macroeconomic data. These give a lower estimate of the numbers in absolute poverty than estimates based on household budget surveys

Table 8.1 Wage costs, productivity and poverty in central and south-east Europe

	Average wage (ECU per month)		Unit labour costs	Output per worker	Real wages	Exchange rate/ PPP ratio 1997	% in poverty 1993–5
	Exchange rate basis	PPP basis	Level in 1997:1993 = 100				
Slovenia	736	956	139	116	134	0.77	<1%
Czech	293	637	126	142	148	0.46	<1%
Poland	283	505	110	173	143	0.56	14%
Hungary	272	432	76	149	104	0.63	2%
Slovakia	250	532	139	119	134	0.47	<1%
Estonia	226	342	264	131	173	0.66	37%
Latvia	193	288	230	151	178	0.67	22%
Lithuania	183	373	407	112	178	0.49	30%
Romania	77	233	90	150	93	0.33	39%
Bulgaria	72	240	58	119	63	0.30	15%

Sources and notes: Monthly wage, unit labour costs and real wages, estimated from data in EBRD (1998), 64–6. ECU wage in Slovenia estimated from OECD Short-Term Economic Indicators 1997/4. Estimate is for 1996. Exchange rate/purchasing power parity ratio estimated from Table 1.1. Poverty headcount measured as percentage of population living on less than $4 per day (ECU 100 per month) on a purchasing power parity basis from Milanovic (1998, p. 75).

where respondents tend to understate actual incomes. A glance at columns 2 and 7 suggests that an average monthly wage of around ECU 400 on a purchasing power parity basis would have been required to reduce absolute levels of poverty to 'reasonable' proportions, if combined with proper targeting of social benefits. Table 8.1 also provides estimates of the growth of unit labour costs in Deutschmarks (DM), output per worker, and real wages over the period from 1993–7. These have been estimated from EBRD data which are based on national sources and need to be treated with some caution. Changes in output per worker in industry frequently reflect reductions in the labour force rather than a growth in output.

Slovenia is a clear outlier with average wages measured in ECU at the prevailing exchange rate that are 2.5 times higher than the next highest country, the Czech Republic. Slovenia also had the highest ratio between income measured on a purchasing power parity basis and on an exchange rate basis of 0.77 which is comparable to that of Portugal (0.76). Monthly wage levels in ECU in the Czech Republic, Poland and Hungary fell within an 8 per cent band. In the majority of central east European economies productivity growth has been accompanied by a growth in real wages and a real appreciation of the exchange rate resulting in a rise in unit labour costs in terms of the DM between 1993 and 1997. The real appreciation of the exchange rate has contributed to a narrowing of the gap between incomes measured on a purchasing power parity basis and an exchange rate basis, but the ratio between the two, still lags considerably behind that recorded by the poorer members of the EU. One outstanding feature of this group is that unit labour costs in Hungary, measured in DM, fell by 24 per cent between 1993 and 1997 as a result of a sustained growth in productivity in manufacturing industry of 49 per cent over the period and a relatively modest growth in real wages. Levels of absolute poverty have remained relatively low in central Europe, as wage rate growth has helped to sustain average earnings in manufacturing above the absolute poverty threshold. Hungary, Slovenia, Slovakia and the Czech Republic, which have the lowest ratios of labour-intensive exports (see Table 6.6), also have the lowest rates of poverty. Poland, where labour-intensive exports goods accounted for 43.1 per cent of exports had a relatively high rate of poverty at 14 per cent. Although long-term unemployment, family size and a large rural sector are major causes of poverty in Poland, 60 per cent of those in poverty in Poland were the working poor, particularly those in unskilled manual and clerical occupations (Milanovic, 1998, 96).

Different patterns apply in the Baltic states and the Balkans. The growth in labour productivity in the Baltic states have been outstripped by rises in real wages, while the employment of fixed exchange-rate regimes has resulted in a substantial real appreciation of exchange rates and increases in unit costs measured in DM. This has contributed to a reduction in the gap between incomes measured in exchange rates and purchasing power parity in Estonia and Latvia, which partly explains why relatively high ECU wages are associated with high rates of absolute poverty on a purchasing power parity basis. Absolute levels of poverty have remained high in the Baltic states since the initial major collapse in output from 1990 to 1993 following the break up of the Soviet Union. Nevertheless, the Baltic states remain highly dependent on labour-intensive exports, of which wage costs are a major component. Poverty in Estonia is highest amongst those outside the industrial labour force including farmers, the unemployed, self-employed and pensioners. However, even in Estonia, a two income family earning 66 per cent per cent of the average wage with three dependants would fall into absolute poverty in the absence of welfare payments.

The position in the Balkan economies is substantially more serious. Both Romania and Bulgaria have experienced major economic problems in the mid 1990s which have resulted from the failure to implement market-oriented reforms and the relaxation of macroeconomic policies at critical stages of the reform process. In both countries centre-right governments, who have attempted to tackle the underlying problems in different ways, were elected in the mid 1990s. In Bulgaria the virtual collapse of the currency resulted in the creation of a currency board regime in 1997 to restore confidence. The new government in Romania allowed the real exchange rate to fall by 25 per cent in early 1997 before allowing it to appreciate in real terms in the remainder of the year. Although the Bulgarian economy had started to stabilise in 1998 after two years of falling GDP and hyperinflationary pressures and had started to attract FDI, the Romanian economy had shown little sign of recovery by mid-1999. This makes it difficult to interpret the range of problems experienced by these economies from the bare data in Table 8.1.

The data in Table 8.1 show that average wages measured in ECU in both countries are less than half that of the next poorest economy, Lithuania, and just over a quarter that of central European economies. This is partly a consequence of the real depreciation of the exchange rate which is reflected in the low ratio of incomes measured on an exchange rate bias to a purchasing power parity basis. This helps to drive up average incomes on a purchasing power parity basis to ECU 233–40 per month. Even then, a family with two workers who both earn average incomes but who support three dependants, would experience absolute levels of poverty in the absence of any other forms of income. The majority of families which have only one employed worker will experience absolute levels of poverty. Under these circumstances the estimates which show that 39 per cent of the population in Romania are in poverty are not surprising.[3] The estimates of poverty in Bulgaria predate the economic crisis and probably underestimate current levels of poverty substantially.

8.5   Conclusion: the implications for EU enlargement

8.5.1   How ready are the central and south-east European economies for membership?

The research in this book indicates that the ten accession states can be divided into three separate structural-economic categories for purposes of enlargement. The first group consists of countries whose trade structures and levels of development, ability to attract foreign direct investment and elimination of absolute poverty suggest that they are capable of withstanding competition within the EU and complying with the requirements of the acquis communautaire with relatively minor adjustments. This group would minimally consist of Slovenia and Hungary and could be extended to include the Czech Republic. These three states are relatively small (with a combined population of 23 million) and are centrally located. It would be expected that these states could become core members of the Union over time and would be capable of participating in moves towards monetary union.

The second group consists of states who would expect to experience some short term difficulties in competing in EU markets, for one reason or another, but may be generally considered to be progressing along an 'upward curve' of economic progress. This group comprises Poland, Slovakia and Estonia and could (just) be extended to include Lithuania. These states remain relatively dependent on labour-intensive exports, while Poland, Estonia and Lithuania still suffer from relatively high levels of poverty. FDI into Estonia is relatively high and it has made some progress in developing exports of human capital-intensive goods which embody medium technology. Slovakia was omitted from the first tier of negotiations for failing to meet political, not economic, criteria. Poland has pursued prudent fiscal policies, is a leader in implementing market-oriented policies, has a relatively high level of private investment and has attracted major inflows of FDI over the last three years. Its economy shows an underlying strength that is not directly affected by changes of government. There are strong grounds for arguing that trade data for 1996 do not fully capture the changes in the economic structure that are currently taking place and that Poland could progress to the first group. However, the size of the country (38.7 million) and the persistence of poverty in some regions, including the rural sector indicate that this will require specific adjustment measures.

The third group consist of Latvia and the two Balkan states, Bulgaria and Romania. These economies have experienced difficulties in implementing economic reforms and consistent sound macroeconomic policies, although Bulgaria implemented a currency board system in 1997 after experiencing a major financial crisis in 1996–97. The export structure of these three countries, with relatively low rates of exports of human capital-intensive goods lags considerably behind the structure of demand inside the EU reflected in intra-EU trade and in EU imports from the rest of the world. Latvia is starting to attract inflows of FDI and relatively small levels of investment would have a significant impact on economic development. Romania is highly dependent on labour-intensive exports and outward processing trade. The level of income in these three countries suggests that it will be some time before poverty is reduced to the levels of the more developed transition economies. Romania and Bulgaria are still experiencing major difficulties in attracting FDI and stabilising production after two consecutive years of falling industrial production. The Balkan economies still face major problems of industrial restructuring, improving infrastructure and developing sound financial systems before they could be considered to be capable of complying with the conditions of the *acquis communautaire*.

8.5.2 The challenge for the EU

It has been indicated above that although a major economic gap exists between the existing members of the EU and the first tier of candidates for accession who opened negotiations for entry in 1999, at least three of those economies (Slovenia, Hungary and the Czech Republic) will be capable of meeting the entry criteria, while the two others (Poland and Estonia) plus Slovakia are making some progress towards achieving levels of economic development that are consistent with membership in the long term. The two remaining Baltic States, Latvia and Lithuania, are relatively small and might be expected to adjust more rapidly as their economies become more closely integrated with those of the Nordic states that are already members of the EU. However the evidence also suggests that the economic gap between the Balkan economies and the central European economies is widening not diminishing and that the economic gap may widen further following the war in Kossovo, which has damaged transport links to central Europe, unless major external assistance is forthcoming. Furthermore there is the danger that the possible exclusion of these economies from the process of enlargement over the next decade will exacerbate the problems by giving an additional impetus to investment to the countries included in the first round of enlargement and creating agglomeration effects and external economies that cannot be replicated by latecomers. Furthermore the economies in the first round of enlargement will benefit from greater EU finance which will assist them in meeting the social costs associated with economic restructuring. The problems of exclusion will also apply to Albania and the states of former Yugoslavia. As the EU enlarges, the costs of exclusion for non-members will actually increase as they face greater problems in trading with former partners and become less attractive to potential investors. At the same time policies that are intended to reduce inequality between existing members of the EU may actually increase inequality between the new members and those who have been excluded on the grounds of their initial backwardness.

The question of how the EU can stimulate economic development in European regions that are not full members of the EU and are only eligible for limited financial assistance raises fundamental questions about the nature of the EU and European security and will create a major challenge for the organisation over the next decade. It is essentially a European question and cannot be simply passed on to the multilateral development agencies. Economic backwardness in the Balkan region results from a combination of the less favourable structural inheritance from communism, the slow pace of economic reform and restructuring and low levels of private sector investment including the relative lack of success in attracting FDI. It is clear that the acceleration of domestic reforms is essential if private sector investment is to be attracted to less developed regions. There is also an urgent need for investment in infrastructure and in education and training. This also raises fundamental questions about why it has proved so difficult to implement reforms in lagging countries. Part of the explanation lies in greater levels of corruption and rent-seeking activity by former elites which have inhibited the development of functioning capital markets and the development of properly-functioning systems of corporate governance and have contributed to the misuse of public revenues at a time when it is essential that public funds should be channelled to the most important areas. However, popular resistance to change and industrial restructuring is not surprising in countries where policies of full employment, overstaffing and subsidisation of basic staple goods were the major method for preventing large numbers of the populations from falling into absolute poverty in the communist era and have not been replaced by properly functioning welfare systems. The potential winners from reform must offer the loser some compensation if they are to win support for measures that are in society's long run benefit. This suggests that EU support for the implementation of welfare policies and retraining and finance and other assistance for starting new businesses for those who lose their major means of support should be extended. The implementation of these measure will require reconsideration of EU budget priorities and will force the EU to reconsider its priorities. This is not simply a question of 'widening' versus 'deepening', but the need to reconsider the best methods of achieving long-term security objectives in Europe following the collapse of communism.

Appendix: CEE and EU Imports

Factor content categories, unit value ratios of EU imports and ratio of CEE:EU imports in EU-15 trade with CEE-10 in 1996

SITC number and description	Factor content category	Extra-EU unit value (ECU per tonne)	Ratio CEE:EU unit value
5	**Chemicals**		
51	**Organic Chemicals**		
511	Hydrocarbons & derivatives	Resource	465	0.688
512	Alcohols, phenols & derivs	Resource	327	1.336
513	Carboxylic acids	Resource	1,552	0.563
514	Nitrogen-function compounds	Resource	3,511	0.350
515	Organo-inorganic compounds	HK-Medium-k	10,210	0.208
516	Other organic chemicals	HK-High-k	1,822	0.343
52	**Inorganic Chemicals**		
522	Inorganic chemical elements	HK-Medium-k	296	0.698
523	Metal salts and peroxysalts	HK-Medium-k	216	0.723
524	Other inorganic chemicals	HK-Medium-k	1,425	0.332
525	Radioactive and associated products	HK-High-k	26,313	0.106
53	**Dyeing tanning and colouring materials**		
531	Synthetic organic colouring	HK-Medium-k	10,157	0.619
532	Dyeing and tanning extracts	HK-Medium-k	1,521	0.558
533	Pigments, paints, varnishes	HK-Medium-k	2,012	0.541
54	**Medicinal and pharmacy**		
541	Other medicinal products	HK-High-k	49,739	0.229
542	Medicaments	HK-High-k	69,154	0.293
55	**Essential oils, resinoids, perfumes**		
551	Essential oils and materials	HK-Medium-k	10,890	0.209
553	Perfumes, cosmetics	HK-Low	9,102	0.282
554	Soaps, cleansing, polishes	HK-Low	10	0.465
56 and 562	**Chemical fertilisers**	Resource	135	0.954

57	**Plastics in primary form**			
571	Polymers of ethylene	HK-Medium-k	709	0.908
572	Polymers of styrene	HK-Medium-k	947	0.825
573	Polymers of vinyl chloride	HK-Medium-k	769	0.652
574	Polyacetals	HK-Medium-k	1,815	0.784
575	Other plastics in primary form	HK-Medium-k	1,760	0.301
579	Waste and scraps	HK-Medium-k	436	0.787
58	**Plastics in non-primary form**	HK-Medium-k		
581	Tubes, pipes, hoses	HK-Medium-k	6,290	0.069
582	Plastic plates, sheets, film, foil	HK-Medium-k	3,711	0.426
583	Monofilaments	HK-Medium-k	4,610	0.471
59	**Other chemical products**			
591	Insecticides, herbicides	HK-High-k	7,580	0.549
592	Starches, albuminoidal	HK-Low	3,175	0.780
593	Explosives, pyrotechnics	HK-Low	2,935	0.860
597	Additives for mineral oils	HK-Medium-k	1,723	0.288
598	Miscellaneous chemicals	HK-Medium-k	1,468	0.350
61	**Leather, leather goods & dressed furs**			
611	Leather	Resource	4,471	1.125
612	Leather manufactures	Labour	7,960	1.889
613	Furskins	Resource	34,780	0.766
62	**Rubber manufactures**			
621	Rubber materials	Labour	3,094	0.535
625	Tyres and inner tubes	Labour	2,851	0.829
629	Rubber articles	Labour	6,413	0.407
63	Cork and wood manufactures			
633	Cork manufactures	Labour	4,356	0.517
634	Veneers, plywood, particle board	Resource	566	0.646

Continued.

Table (*Contd*)

SITC number and description	Factor content category	Extra-EU unit value (ECU per tonne)	Ratio CEE:EU unit value
635 Wood manufactures	Resource	879	0.581
64 Paper, paperboard and articles			
641 Paper and paperboard	Resource	645	0.872
642 Articles of paper and paperboard	Labour	1,921	0.513
65 Textile yarn and fabrics			
651 Textile yarn	Labour	2,891	0.991
652 Cotton fabrics, woven	Resource	4,605	1.142
653 Fabrics of man-made textiles	Resource	6,287	0.953
654 Other textile fabrics, woven	Resource	4,716	1.019
655 Knitted or crocheted fabrics	Labour	6,560	1.188
656 Tulles, lace, embroidery,	Labour	16,013	0.534
657 Special yarns and fabrics	Labour	5,813	0.426
658 Made-up textile articles	Labour	5,380	1.160
659 Floor coverings and carpets	Resource	7,014	0.653
66 Non-metallic mineral manufactures			
661 Lime, cement, construction matls	Resource	58	0.789
662 Clay construction materials	Resource	344	0.679
663 Mineral manufactures	Resource	751	0.299
664 Glass	Resource	872	0.658
665 Glassware	Resource	1,336	0.737
666 Pottery	Resource	1,747	1.184
667 Precious metals, stones, pearls	Labour	2,533,926	0.087
67 Iron and Steel			
671 Pig iron, granules and ferro-alloys	Resource	425	0.871

672	Ingots, primary and semi-finished products of iron and steel	Resource	275	0.914
673	Flat-rolled, non-alloy, unclad	Resource	292	0.982
674	Flat-rolled, non-alloy, clad	Resource	585	0.745
675	Flat-rolled products of alloy steel	Resource	1,246	0.854
676	Iron and steel bars, rods, shapes	Resource	396	0.770
677	Rails and railway track	Resource	424	0.991
678	Wire of iron and steel	Resource	647	0.562
679	Tubes, pipes, hollow fittings	HK-Medium-k	792	0.714
68	**Non-ferrous metals**			
681	Silver and platinum	Resource	214,769	0.632
682	Copper	Resource	1,975	0.953
683	Nickel	Resource	6,839	1.137
684	Aluminium	Resource	1,599	0.942
685	Lead	Resource	761	0.854
686	Zinc	Resource	917	1.067
687	Tin	Resource	4,890	1.008
689	Miscellaneous non-ferrous metals	HK Medium-l	2,065	2,972
69	**Manufactures of metals**			
691	Structures and parts of structures of iron, steel and aluminium	HK Medium-l	1,384	0.775
692	Metal storage containers	Labour	2,266	0.645
693	Wire products, fencing	Resource	1,005	0.433
694	Nails, screws, nuts, bolts etc.	Resource	2,314	0.436
695	Tools for use by hand or machine	H-K Medium-l	8,548	0.600
696	Cutlery	Labour	9,543	0.771
697	Household goods of base metal	Labour	3,704	0.531
699	Manufactures of base metal	Labour	2,726	0.531

Continued.

Table (*Contd*)

SITC number and description	Factor content category	Extra-EU unit value (ECU per tonne)	Ratio CEE:EU unit value
7 Machinery and equipment			
71 Power generating			
711 Steam and other boilers	HK-Low	3,705	0.695
712 Steam turbines	HK-Low	22,613	0.384
713 Internal combustion engines	HK-Low	8,705	0.895
714 Engines and motors	HK-High-l	294,279	0.093
716 Rotating electric plant & motors	HK-Low	8,306	0.493
718 Other power generating equipmt	HK-High-l	14,369	0.362
72 Specialised by industry			
721 Agricultural machinery	Labour	4,312	0.525
722 Tractors	Labour	4,904	0.772
723 Civil engineering equipment	HK Medium-l	4,001	0.416
724 Textile and leather machinery	Labour	13,200	0.397
725 Paper-making machinery	HK Medium-l	13,310	0.253
726 Printing, bookbinding equipment	HK Medium-l	23,790	0.191
727 Food processing equipment	HK Medium-l	14,128	0.445
728 Other specialised equipment	HK Medium-l	15,162	0.232
73 Metal working machinery			
731 Machine tools for remving metal	HK Medium-l	14,959	0.345
733 Machine tools for working metal	HK Medium-l	9,140	0.349
735 Parts and accessories for 731,733	HK Medium-l	12,590	0.246
737 Other metalworking machinery	HK Medium-l	9,462	0.321
74 General industrial machinery			
741 Heating and cooling equipment	HK Medium-l	12,006	0.352
742 Pumps for liquid	HK-Low	14,441	0.531

743	Pumps not for liquid	HK-Low	8,801	0.398
744	Mechanical handling equipment	HK Medium-l	4,523	0.418
745	Non-electrical machinery	HK Medium-l	14,822	0.293
746	Ball bearings	HK Medium-l	8,287	0.623
747	Taps, cocks, valves	HK Medium-l	12,295	0.314
748	Transmission shafts	HK Medium-l	5,703	0.469
749	Non-electrical parts	HK Medium-l	14,180	0.397
75	**Office machines and computers**			
751	Office machines	HK Medium-l	24,089	0.330
752	Computers	HK-High-k	46,823	0.556
759	Parts and accessories for 751–2	HK Medium-l	51,895	0.445
76	**Telecommunications & recording**			
761	Television receivers	HK Medium-l	13,890	0.769
762	Radio receivers	HK-Low	15,193	0.888
763	Sound recording apparatus	HK Medium-l	36,210	0.606
764	Telecommunications equipment	HK Medium-l	39,416	0.477
77	**Electrical machinery**			
771	Electric power machinery	Labour	14,279	0.700
772	Electrical apparatus, circuits	HK Medium-l	34,789	0.396
773	Equip for distributing electricity	HK Low	7,746	0.748
774	Electro-diagnostic apparatus	HK Medium-l	95,019	0.269
775	Household electrical equipment	HK Medium-l	5,545	0.664
776	Valves and transistors	HK High-l	91,882	0.083
778	Electrical machinery nes	HK High-l	17,091	0.450
78	**Road vehicles**			
781	Passenger cars	HK Medium-k	7,551	0.865
782	Vehicles for transport of goods and special purpose vehicles	HK Medium-k	6,580	1.143
783	Other road vehicles	HK Medium-k	7,719	0.588

Continued

Table (*Contd*)

SITC number and description	Factor content category	Extra-EU unit value (ECU per tonne)	Ratio CEE:EU unit value
784 Parts for road vehicles	Labour	5,809	0.630
785 Motorbikes and cycles	Labour	10,748	0.357
786 Trailers, caravans	Labour	1,403	0.958
79 Other transport equipment			
791 Railway vehicles, locomotives	Labour	3,395	0.495
792 Aircraft and equipment	HK High-l	372,548	0.183
793 Ships and boats	HK Low	3,770	0.508
81 Prefabricated buildings, lighting, fittings			
811 Prefabricated buildings	HK Medium-l	1,064	0.844
812 Plumbing and heating fittings	Labour	1,962	0.640
813 Lighting fixtures and fittings	Labour	5,049	0.710
82 and 821 Furniture and bedding	Labour	2,570	0.795
83 and 831 Travel goods, bags & cases	Labour	7,213	0.979
84 Clothing and apparel	**Labour**		
841 Male clothing, not crocheted	Labour	16,215	1.433
842 Female clothing, not crocheted	Labour	21,630	1.283
843 Male clothing, crocheted	Labour	12,686	1.088
844 Female clothing, crocheted	Labour	14,285	1.229
845 Garments made up from textile	Labour	15,807	1.149
846 Clothing accessories from textile	Labour	13,920	1.168
848 Clothing & accessories not from textile fabrics	Labour	10,785	1.753
85 and 851 Footwear	Labour	10,704	1.281
87 Professional, scientific and controlling instruments			
871 Optical instruments and apparatus	HK High-l	73,852	0.844

199

Code	Description		Class	
872	Instruments and appliances, medical, surgical, dental and veterinary	44,333	HK Medium-l	0.163
873	Meters and counters	33,581	HK Medium-l	0.447
874	Measuring,control instruments	92,217	HK High-l	0.190
88	**Photographic apparatus, optical goods, clocks and watches**			
881	Photographic apparatus	74,733	HK Medium-l	0.101
882	Photographic, cinematic supplies	20,245	HK Medium-k	0.623
884	Optical goods	76,714	HK Medium-l	6.811
885	Watches and clocks	74,324	Labour	1.054
89	**Miscellaneous manufactures**			
891	Arms and ammunition	37,097	Labour	0.224
892	Printed matter	5,954	Labour	0.420
893	Plastic articles nes	3,110	Labour	0.751
894	Toys	6,761	Labour	0.749
895	Office and stationery supplies	10,933	Labour	0.410
897	Jewellery and articles of precious metals	196,314	Labour	0.361
898	Musical instruments	23,903	H-K-Low	0.452
899	Miscellaneous manufactures	7,693	Labour	0.293

Notes and sources: UN (1994). SITC Classifications from UN Commodity Index for the Standard International Trade Classification, Revision 3. These have been abbreviated for purposes of clarity and brevity. Factor Content categories are derived from the Legler–Schulmeister system adapted by Wolfmayr-Schnitzer (1998) and are described in Chapters 3 and 6.

Key:
HK High-l: Human capital-intensive, embodying high technology and labour-intensive processes.
HK High-k: Human capital-intensive, embodying high technology and capital-intensive processes.
HK Medium-k: Human capital-intensive, embodying medium technology and capital-intensive processes.
HK Medium-l: Human capital-intensive, embodying medium technology and labour-intensive processes.
HK-Low: Human capital-intensive, embodying low technology.
Labour: Labour-intensive.
Resource: Resource-intensive.

Notes

Chapter 1

1 These are Armenia, Azerabaijan, Belarus, Georgia, Kazakhstan, Kyrgyzstan, Moldova, Russia, Tajikistan, Turkmenistan, Ukraine and Uzbekistan.

2 The Baltic States fell under Russian Tsarist rule in the eighteenth century and gained a brief period of independence in the inter-war period before being incorporated into the Soviet Union as a result of the Nazi–Soviet pact of 1939.

3 These agreements were subsequently changed into Europe Agreements in 1994 (Poland and Hungary) and 1995 (Romania, Bulgaria, Czech Republic and Slovakia). The Baltic states concluded Europe Agreements in 1995 and Slovenia in 1996. Europe Agreements, which create more open trade relations, do not carry a guarantee of eventual accession.

4 Sensitive goods are goods whose importation into the EU is considered to be 'sensitive' from the perspective of unemployment in the EU. These are essentially labour-intensive goods with a low income elasticity of demand and for which overcapacity exists in the EU.

5 These were: 1973, Denmark, Ireland, UK; 1981, Greece; 1986, Portugal and Spain; 1995, Austria, Finland, Sweden.

6 For example, the inclusion of Hungary in the first tier of accession will require it to impose visa restrictions on ethnic Hungarians living in Romania and Slovakia who have not been included in the first tier of accession.

7 Despite the high volume of Finland's exports to the USSR, exports to the EU-15 still constituted nearly two-thirds of Finland's exports in 1989.

8 For purposes of comparison the USA, which conducts a high volume of trade within its own borders was responsible for 16 per cent of world trade flows in 1989.

9 Estimates of incomes on a market basis for the EU economies have been estimated by the World Bank using a methodology that takes account of changes in incomes, exchange rates and differential inflation over a three-year period to eliminate the effect of short-run movements in these variables.

10 The major purpose of the single market and monetary union is to reduce transactions costs between member states to stimulate intra-EU trade.

11 If two countries have an identical GDP, but different population sizes, we would expect the country with the higher GDP per-capita to trade more than the country with a smaller GDP per capita.

12 The accession economies have been arranged geographically in this and subsequent tables with the central European states arranged according to the size of trade turnover with the EU, followed by the Balkan states and the Baltic states.

13 This form of massive rupture from the past to break away from inherited backwardness is exactly what central planners attempted to create in eastern Europe in the initial five-year plans in the 1950s.

Chapter 2

1 For a cogent criticism of the argument that nations cannot be viewed as 'competitive entities' like football teams, whereby the employment prospects and the standard of living of their citizens depends on its relative standing and competitive status in international markets see Krugman (1994, 1996).

2 The factor price equalisation theorem was first developed by Samuelson (1948). Assuming the absence of economies of scale and factor reversals, identical international technology and zero transport costs, it can be demonstrated that trade liberalisation will lead to the complete equalisation of factor prices under conditions of perfect competition.

3 This raises a major question about whether the inherited labour forces are intensive in human skills. Halpern (1995) makes a strong case to suggest that participation rates for primary, secondary and tertiary education in the majority of CEE economies under communism were high in comparison with those of traditional labour-intensive and capital-scarce economies. However, much of the training received was industry-specific and did not provide the skills that are required in a new economic environment.

4 Whether this is 'good' or 'bad' for the population of the country from which people emigrate depends on how the population is defined. Policies which promote emigration which leads to higher incomes and welfare for emigrants will result in increased aggregate economic welfare for the *existing* population (including émigrés) of the country concerned. However the emigration of workers who earn above-average wages will result in a lower observed GDP per capita for those that *remain* in the country, even though they as individuals may not be any worse off. However, the emigration of high-income earners could contribute to a lower tax-base and difficulties in financing expenditure on infrastructure in the home country.

5 Krugman (1991, ch. 2) stresses that industrial concentration in the US does not just affect sectors involving high-technology but affects sectors like the carpet industry, the shoe industry and the automobile industry.

6 These studies are based on the standard measure for estimating intra-industry trade developed by Grubel and Lloyd (1975).

Chapter 3

1 The categories used by Wolfmayr-Schnitzer have been modified by eliminating a category of physical capital-intensive, which accounted for 2–4 per cent of CEE exports to the EU. These exports have been re-allocated to either resource-intensive or labour-intensive categories. I have also made some other minor re-allocations of products between categories and have omitted data on trade in all categories of SITC 9 (articles not classified elsewhere) from the estimates. Wolfmayr-Schnitzer describes the category of low technology as 'others'.

2 The export specialisation index was first developed by Hisao Kanamori (1964), cited by Vollrath (1991). Gual and Martin (1995) have applied this test to compare Spain's trade with the EU with that of CEE economies. Dimelis and Gatsios (1995) have applied the RCA test to trade between Greece and five CEE economies.

3   This will apply to all industrial sectors in which CEE-10 exports to the EU were between 68 per cent and 100 per cent of CEE imports from the EU in the tables in this chapter derived from 1996 data as total CEE-10 exports of manufactured goods to the EU were only 68 per cent of CEE-10 imports of manufactured goods.

4   For example, an index of four in industrial sector i for a given country and an index of two in industrial sector j cannot be taken to prove that the exporting country is twice as competitive in sector i as in sector j, or even that the exporting country is *more* competitive in sector i than in sector j. Purely mathematically, the ESI index could not provide a cardinal measure of comparative advantage between different goods unless the indexes were reduced to logarithms.

5   An ESI index of 3 for sector i for country X and 1.5 for country Y *does* indicate that the exports from sector i were twice as important for country X as for country Y.

Chapter 4

1   These consisted of Bulgaria, Czechoslovakia, German Democratic Republic (GDR) Hungary, Poland and Romania.

2   Although these proportions may be biased upwards by the relative overvaluation of machinery and equipment in intra-CMEA prices, unspecified trade in which machinery and equipment is believed to predominate, is only included in the denominator.

3   Commodity balances provide a clearer picture of the pattern of trade than indices of revealed comparative advantage (RCA) described in Chapter 2. For example, Bulgaria did not export any energy or wood and paper products to the USSR in 1988 which would have resulted in an RCA index of −1 for both categories, although the deficit in fuels and energy is clearly more significant.

4   Aggregated figures for these items are specified in the Soviet import data from each country except Bulgaria. Consequently if trade in these items did occur, but was not recorded in the Soviet data, this would represent deliberate omission rather than underreporting. It is not clear why the Soviet authorities would choose to do this for an area of trade with no obvious strategic significance.

5   The reform economist and deputy-premier, Ota Sik, was a leading exponent of this argument during the Czechoslovak reforms of 1967–8. See Sik (1971). See also Abonyi (1981) for the links between the New Economic Mechanism in Hungary and imported technology.

Chapter 5

1   A thorough analysis of the debt-cycle is given in Williamson and Milner (1991), ch. 16.

2   Trade with countries outside the EU accounted for 37 per cent of the total exports of the EU-15 economies (including intra-EU trade) in 1996 and for 35.4 per cent of their imports.

3 Textile yarns, which are largely imported under outward processing agreements for the manufacture of clothing which are re-exported (see below) have been included as items of consumption under imports. This gives a clearer indication of net imports of consumer goods for domestic consumption.
4 OPT arrangements can involve more than one EU partner when the subcontractor delivers the good to a customer in a third country. For example, Italian yarns may be delivered to Romania for cutting and trimming and re-exported to the UK.
5 This will be the case as a matter of statistical reporting, although subcontracting may persist.
6 Estimates of the UVIs for EU imports from the World and and the ratios of EU import UVIs from CEE economies to UVIs of EU imports from the world for all manufactured goods at the three digit levels of the SITC classification are also shown in the Appendix.
7 The ratio is given as

$$\text{Ratio of UVIs} = \frac{\text{UVIi CEE}}{\text{UVIi World}} = \frac{\text{Vi CEE/Vi World}}{\text{Wi CEE/WiWorld}}$$

Where UVIi = unit value of imports of good i
i = imports of good i
Vi = value of imports of good i by EU in thousand ECU
Wi = weight of imports of good i in metric tonnes
CEE = imports from central and east Europe
world = imports from world

8 A lower index at a higher level of aggregation reflects the relative inability to penetrate high value markets within that product category.

Chapter 6

1 A controversial example is the decline of coal mining in the UK.
2 Distortions arising from the enlargement of the EU to absorb the eastern Lander of Germany cannot be avoided. 1994 also is the year in which all of the former CMEA economies recorded positive growth and therefore could be considered to be emerging from the initial transition recession and provides some clear information on the history of the initial stage of the transition.
3 The slow growth of EU imports of manufactured goods from Romania in this period cannot be attributed to excessive Romanian exports in the Ceausescu era which were largely concentrated on fuels and energy and foodstuffs.

Chapter 7

1 Industries in which the CEE economies generate a positive RCA index but an ESI of less than one, may be areas where it trade liberalisation may be expected to result in trade diversion (that is replacing EU imports from outside the EU by imports from the CEE).
2 The value of exports of ships from individual countries to the EU is highly volatile from year to year as a result of the high value of a single delivery which

embodies several months production. Poland was the sixth largest exporter of ships and boats to the EU in 1993. Similarly the destination of exports of ships is affected by the registration of vessels in third countries.

Chapter   8

1 Trade diversion includes the destruction of efficient trade links between the CEE economies and trade partners outside the EU (particularly in the CIS) as a result of the adoption of common external tariffs on imports from outside the EU.
2 Data and comment in this section are based on the work of Milanovic (1998, 60–113). Any errors of interpretation are, of course, my responsibility.
3 This proportion rises to 59 per cent if data based on household budget surveys are used instead of macroeconomic data (Milanovic, 1998, 68).

Bibliography

Abonyi, A. (1981). 'Imported Technology, Hungarian Industrial Development and Factors Impeding the Emergence of Innovative Capacity', in P.G. Hare, H.K. Radice, N. Swain (eds), *Hungary: A Decade of Reform*. London: George Allen and Unwin.

Aiginger, Karl (1998). 'Unit Values to Signal the Quality Position of CEECs', in *The Competitiveness of Transition Economies*. OECD Proceedings: 15–40.

Aturupane, Chonira, Simeon Djankov and Bernard Hoekman (1997). Determinants of Intra-Industry Trade between East and West Europe. *World Bank Discussion Paper*.

Avery, Graham and Fraser Cameron (1999). *The Enlargement of the European Union*. Sheffield: Academic Press.

Balassa, Bela (1965). 'Trade Liberalisation and Revealed Comparative Advantage', *The Manchester School of Economic and Social Studies*, 33: 99–123.

Baldwin, Richard (1994). *Towards an Integrated Europe*. London: Centre for Economic Policy Research.

Berend, Istvan (1971). 'The Problem of East European Integration in Historical Perspective', in I. Vajda and M. Simai (eds), *Foreign Trade in a Planned Economy*. Cambridge: University Press.

Bianchini, Stefano and Milica Uvalic (eds) (1997). *The Balkans and the Challenge of Economic Integration*. Ravenna: Longo Editore.

Brenton, Paul and Daniel Gros (1995). *Trade between the European Union and Central Europe, an Economic Policy Analysis*. Working Document 93. Brussels: Centre for European Policy Studies.

Brenton, Paul and Daniel Gros (1997). 'Trade Reorientation and Recovery in Transition Economies', *Oxford Review of Economic Policy*, 13 (2): 65–6.

Carlin, Wendy and Michael Landesmann (1997). 'From Theory into Practice? Restructuring and Dynamism in Transition Economies', *Oxford Review of Economic Policy*, 13 (2): 77–105.

Cooper, Richard and Janos Gacs (1997). 'Impediments to Exports in Small Transition Economies', *Most-Most*, 7 (2): 5–32.

Centre for Finnish Business and Policy Studies (1997). *More Members for the EU?* Helsinki: EVA.

CEPII – European Commission (1997). *Trade Patterns inside the Single Market: Single Market Review. Subseries IV: vol. 2*. Luxembourg: European Communities.

Dobrinsky, Rumen and Michael Landesmann (eds) (1995). *Transforming Economies and European Integration*. Cheltenham: Edward Elgar.

Dobrinsky, Rumen and Ira Yaneva (1997). 'Impediments to Exports in Small Transition Economies: The Case of Bulgaria', *Most-Most*, 7 (2): 32–55.

Eatwell, John, Michael Ellman, Mats Karlsson, Mario Nuti and Judith Schapiro (1997). *Not Just Another Accession – The Political Economy of Enlargement to the East*. London: Institute for Public Policy Research.

EBRD (1995). *Transition Report 1995*. London: EBRD.

EBRD (1996). *Transition Report 1996*. London: EBRD.

EBRD (1997). *Transition Report 1997*. London: EBRD.

EBRD (1998). *Transition Report 1998*. London: EBRD.

EBRD (1999). *Transition Report Update 1999*. London: EBRD.

Ellingstad, Marc (1997). *The Maquiladora Syndrome: Central European Prospects. Europe-Asia Studies*, 49 (1): 7–23.

Estrin, Saul, Kirsty Hughes and Sarah Todd (1997). *Foreign Direct Investment in Central and Eastern Europe: Multinationals in Transition*. London: Royal Institute of International Affairs/Pinter.

Falvey, Rodney (1981). 'Commercial Policy and Intra-industry Trade', *Journal of International Economics*, 11: 495–511.

Feldman, R., K. Nashashibi, R. Nord, P. Allum, D. Desruelle, K. Enders, R. Kahn, and H. Temprano-Arroyo (1998). *Impact of EMU on Selected Non-European Union Countries*. Washington, DC: IMF.

Faini, Riccardo and Richard Portes (eds) (1995). *European Union Trade with Eastern Europe: Adjustment and Opportunities*. London: Center for Economic Policy Research.

Ferreira, M.P. (1995). 'The Liberalisation of East–West Trade: An Assessment of its Impact on Exports from Central and Eastern Europe', *Europe-Asia Studies*, 47 (7): 1205–23.

Gatsios, Konstantine and Sophia Dimelis (1995). 'Trade with Central and Eastern Europe: The Case of Greece', in R . Faini and R. Portes (eds), *European Union Trade with Eastern Europe: Adjustment and Opportunities*. London: Centre for Economic Policy Research.

Gual, Jordi and Carmela Martin (1995). 'Trade and Foreign Direct Investment with Central and Eastern Europe: Its Impact on Spain', in R. Faini and R. Portes (eds), *European Union Trade with Eastern Europe: Adjustment and Opportunities*. London: Centre for Economic Policy Research.

Grabbe, Heather and Kirsty Hughes (1997). *Eastward Enlargement of the European Union*. London: Royal Institute of International Affairs.

Greenaway, David and Chris Milner (1985). *The Economics of Intra-Industry Trade*. Oxford: Blackwell.

Grossman, G.M. and E. Helpman (1991). 'Quality ladders and Product Cycles', *Quarterly Journal of Economics*, 106 (2): 557–86.

Grubel, Herbert and P.J. Lloyd (1975). *Intra-industry Trade: The Theory and Measurement of International Trade in Differentiated Products*. London: Macmillan.

Helpman, Elhanan (1981). 'International Trade in the Presence of Product Differentiation, Economies of Scale and Monopolistic Competition: A Chamberlin–Heckscher–Ohlin Approach', *Journal of International Economics*, 11: 305–40.

Helpman, Elhanan and Paul Krugman (1985). *Market Structure and Foreign Trade: Increasing Returns, Imperfect Competition and the International Economy*. Cambridge, Mass.: MIT Press.

Halpern, Laszlo (1995). 'Comparative Advantage and Likely Trade Patterns of the CEECs', in R. Faini and R. Portes (eds), *European Union Trade with Eastern Europe: Adjustment and Opportunities*. London: Centre for Economic Policy Research.

Hamilton, Carl and Alan Winters (1992). 'Opening up International trade with Eastern Europe', *Economic Policy*, 14 (April): 78–104.

Hirschman, A.D. (1975). *The Strategy of Economic Development*. New Haven: Yale University Press.

Holzman, Franklin (1979). 'Some Systemic Factors Contributing to the Convertible Currency Shortages of CPEs', *American Economic Review*, 69.

Hunya, Gabor (1997). 'Large Privatisations, Restructuring and Foreign Direct Investment', in Z. Salvatore (ed.), *Lessons from the Economic Transition: Central and Eastern Europe in the 1990s*. London: Kluwer.

Hunya, Gabor (1998). 'Recent Developments of FDI and Privatisation', *The Vienna Institute Monthly Report*, 5: 1–7.

Johnson, Harry (1975). *Technology and Economic Interdependence*. London: Macmillan.

Kaminski, Bartolomiej (1998). 'Poland's Transition form the Perspective of Performance in EU Markets', *Communist Economies and Economic Transformation*, 10 (4): 217–40.

Kaminski, Bartolomiej, Zhen Kun Wang and Alan Winters (1996). 'Export Performance in Transition Economies', *Economic Policy*, 23: 423–42.

Kanamori, Hisao (1964). *Exports of Manufactures and Industrial Development of Japan*. Geneva: UN Document E/Conf.46.

Krugman, Paul (1981). 'Intra-industry Specialization and the Gains from Trade', *Journal of Political Economy*, 89 (5): 950–73.

Krugman, Paul (1991). *Geography and Trade*. Leuven and Cambridge, Mass.: Leuven University Press and MIT University Press.

Krugman, Paul (1994). 'Competitiveness: A Dangerous Obsession', *Foreign Affairs*, 73 (2): 28–44.

Krugman, Paul (1996). *Pop Internationalism*. Cambridge, Mass: MIT Press.

Krugman Paul and Anthony Venables (1990). 'Integration and Competitiveness of Peripheral Industry', in C. Bliss and J. Braga de Macedo (eds), *Unity with Diversity in the European Community*. Cambridge: University Press.

Lancaster, K. (1980). 'Intra-Industry Trade Under Perfect Monopolistic Competition', *Journal of International Economics*, 10 (2): 151–75.

Landesmann, Michael and Johann Burgstaller (1998). 'Vertical Product Differentiation in EU Markets: The Relative Position of East European Producers', in *The Competitiveness of Transition Economies*. OECD Proceedings: 123–58.

Landesmann, Michael and Istvan Szekely (1995). *Industrial Restructuring and Trade Re-orientation in Eastern Europe*. Cambridge: University Press.

Lankes, Hans-Peter and Tony Venables (1996). 'Foreign Direct Investment in Economic Transition: The Changing Pattern of Investment', *Economics of Transition*, 4 (2): 331–47.

League of Nations (1942). *The Network of World Trade*. Geneva: League of Nations.

Leamer, Edward (1984). *Sources of International Comparative Advantage: Theory and Evidence*. Cambridge, Mass.: MIT Press.

Legler, H. (1982). 'Zur Position der Bundesrepublik Deutschland im Internationalen Wettbewerb', *Forschungsberichte des Niedersachsischen Instituts fur Wirtschaftsforschung*, 3.

Liontief, Wassily (1953). 'Domestic Production and Foreign Trade: The American Capital Position Re-examined', *Proceedings of the American Philosophical Society*, 97: 331–49.

Martin, Roderick (1998). 'Central and Eastern Europe and the International Economy: The Limits to Globalisation', *Europe-Asia Studies*, 50 (1): 7–26.

Milanovic, Branko (1998). *Income, Inequality and Poverty during the Transition from Planned to Market Economy*. Washington, DC: The World Bank.

Meyer, Klaus (1998). *Direct Investment in Economies in Transition*. Cheltenham: Edward Elgar.

Micklewright, John (1999). 'Education, Inequality and Transition', *The Economics of Transition*, 7 (2): 343–75.

Neven, Damien (1995). 'Trade Liberalisation with Eastern Nations: How Sensitive?', in R. Faini and R. Portes (eds), *European Union Trade with Eastern Europe: Adjustment and Opportunities*. London: Centre for Economic Policy Research.

Ohlin, Bertil (1933). *Interregional and International Trade*. Cambridge, Mass.: Harvard University Press.

Porter, Michael (1990). *The Competitive Advantage of Nations*. New York: Free Press.

Posner, Michael (1961). 'International Trade and Technical Change', *Oxford Economic Papers*, 13: 323–41.

Prosi, Gerhard (1998). 'Economic Cooperation between members of the European Union and the New Democratic Countries in Europe', *Communist Economies and Economic Transformation*, 10 (1): 111–18.

Radosevic, Slavo (1999). *International Technology Transfer and Catch-up in Economic Development*. Cheltenham: Edward Elgar.

Rollo, Jim and Alasdair Smith (1993). 'The Political Economy of Eastern European Trade with the European Community: Why so sensitive?', *Economic Policy*, 16 (April): 139–81.

Samuelson, Paul (1948). 'International Trade and the Equalisation of Factor Prices', *Economic Journal*, 58: 163–84.

Schulmeister, S. (1990). 'Das technologische Profil des Österreichischen Aussenhandels', *WIFO-Monatsberichte*, 63 (12): 663–75.

Seliger, Bernhard (1998). 'Integration of the Baltic States in the European Union in the Light of the Theory of Institutional Competition', *Communist Economies and Economic Transformation*, 10 (1): 95–110.

Sen, Amartya (1995). *Mortality as an Indicator of Economic Success and Failure*. Innocenti Lectures. Florence: UNICEF.

Sheets, Nathan and Simona Boata (1996). *Eastern European Export Performance During the Transition*. International Finance Discussion Paper no. 562. Washington, DC: Board of Governors of the Federal Reserve.

Sik, Ota (1971). *Czechoslovakia: The Bureaucratic Economy*. White Plains, NY: IASP.

Smith, Alan (1983). *The Planned Economies of Eastern Europe*. London: Croom Helm.

Smith, Alan (1993). *Russia and the World Economy: Problems of Integration*. London: Routledge.

Smith, Alan (1994). *International Trade and Payments in the Former Soviet/CMEA Area*. London: Royal Institute for International Affairs.

Smith, Alan (1998). 'Trading Places: Changing Trade Patterns between Transition Economies and the EU', *Economies in Transition*, 3: 5–17.

Stern, Robert and Keith Maskus (1981). 'Determinants of the Structure of US Foreign Trade', *Journal of International Economics*, 11: 207–24.

Stolper, Wolfgang and Paul Samuelson (1941). 'Protection and Real Wages', *Review of Economic Studies*, 9: 58–73.

Stolze, Frank (1997). 'Changing Foreign Trade Patterns in Post-reform Czech Industry (1989–95)', *Europe-Asia Studies*, 49 (7): 1209–36.

Tarr, David (1992). 'Problems in the Transition from the CMEA: Implications for Eastern Europe', *Communist Economies and Economic Transformation*, 4 (1): 23–43.

UNCTAD (1998). *World Investment Report 1998: Trends and Determinants*. New York and Geneva: UN.

UNECE (1992). *Economic Bulletin For Europe*, vol. 44. New York Geneva: UN.

UNECE (1998). *Economic Survey of Europe 1998*, no. 1. New York and Geneva: UN.

Wang, Zhen Kun and Alan Winters (1992). 'The Trading Potential of Eastern Europe', *Journal of Economic Integration*, 7: 113–36.

Wiles, Peter and Alan Smith (1978). 'The Convergence of the CMEA on the EEC', in A. Shlaim and G.N. Yannopoulos (eds), *The EEC and Eastern Europe*. Cambridge: University Press.

Van Brabant, Jan (1980). *Socialist Economic Integration*. Cambridge: University Press.

Vernon, Raymond (1966). 'International Investment and International Trade in the Product Cycle', *Quarterly Journal of Economics*, 80: 121–5.

Vollrath, Thomas (1991). 'A Theoretical Evaluation of Alternative Trade Intensity Measures of Revealed Comparative Advantage', *Weltwirtschaftliches Archiv*, 127 (2): 265–79.

Williamson, John and Chris Milner (1991). *The World Economy*. London: Harvester Wheatsheaf.

Wolfmayr-Schnitzer, Yvonne (1998). 'Trade Performance of CEECs According to Technology Classes', in *The Competitiveness of Transition Economies*. Paris: OECD Proceedings: 41–70.

Wolfmayr-Schnitzer, Yvonne (1998b). 'Intra-Industry Trade of CEECs', in *The Competitiveness of Transition Economies*. Paris: OECD Proceedings: 81–92.

World Bank (1999). *World Development Report: Knowledge for Development*. New York: Oxford University Press.

Yeats, Alexander (1985). 'On the Appropriate Interpretation of the Revealed Comparative Advantage Index: Implications of a Methodology Base on Industry Sector Analysis', *Weltwirtschaftliches Archiv*, 121: 61–73.

Zloch-Christy, Iliana (1988). *Debt Problems of Eastern Europe*. Cambridge: University Press.

Statistical sources referred to in the text

Eurostat. Comext Database.

Eurostat (1996). *External and Intra-European Trade. Statistical Yearbook 1958–94*. Luxembourg.

Eurostat (1997). *External and Intra-European Trade. Statistical Yearbook 1958–95*. Luxembourg.

Eurostat (1997a). *External and Intra-European Trade. Statistical Yearbook 1958–96*. Luxembourg.

IMF (1998). *Direction of Trade Yearbook 1996 (1989–95)*. Washington, DC: IMF

IMF (1998). *Direction of Trade Yearbook 1998 (1991–97)*. Washington, DC: IMF

IMF. International Financial Statistics. Monthly. Washington, DC.

OECD. *Short-term Economic Indicators. Transition Economies*. Quarterly, 1992–7. Paris: OECD.

UN (1994). *Commodity Indexes for the Standard International Trade Classification. Revision 3*. Statistical Papers. Series M no. 38/ Rev. 2, vol. 11. Department for

Economic and Social Information and Policy and Statistical Division. New York: United Nations.

National data sources (annual, unless stated)

Bulgaria: *V'nshna T'rgovliya na Republika B'lgariya*. Nationalen Statisticheski Institut, Sofia.

Estonia Latvia, Lithuania, Foreign Trade 1997. Riga: Central Statistical Bureau of Latvia.

Romania: *Anuarul Statistic al Romaniei*. Comisia Nationala pentru Statistica, Bucharest.

CMEA: *Statisticheskii Yezhegodnik Stran-Chlenov*. SEV, Moscow.

USSR: *Vneshniye Ekonomicheskiye Svyazi CCCR v 1988 (1989)*. Statisticheckii Sbornik, Moscow.

Index

References to individual central and south-east European economies and authors cited in brackets in the text have not been included in the index. This book contains an unusually large volume of data presented in tables. Readers may find that the list of tables on pages viii–xi provides a useful guide to the location and analysis of this data.